Supramolecular Structure and Function 10

Jasminka Brnjas-Kraljević · Greta Pifat-Mrzljak
Editors

Supramolecular Structure and Function 10

Springer

Editors
Jasminka Brnjas-Kraljević
School of Medicine
University of Zagreb
10000 Zagreb
Croatia
kraljevi@mef.hr

Greta Pifat-Mrzljak†
Ruđer Bošković Institute
10000 Zagreb
Croatia
pifat@irb.hr

ISBN 978-94-017-8242-5
ISBN 978-94-007-0893-8 (eBook)
DOI 10.1007/978-94-007-0893-8
Springer Dordrecht Heidelberg London New York

Softcover re-print of the Hardcover 1st edition 2011

Printed on acid-free paper

Springer is part of Springer Science+Business Media (www.springer.com)

Preface

In the fifties of the last century the definition of biophysics arose much dispute among the scientists who were traditional physicists, chemists and biologists by training. As an interdisciplinary science, biophysics shares significant overlap with biochemistry, bioengineering and systems biology, but on the other hand offers a rational language for discussion about a common subject to scientists of different scientific disciplines. Biophysics has gradually erased the frontiers in scientific research by bringing together scientists from different fields of research. Nowadays, it has been widely accepted that the search for new knowledge depends not only on new methods and concepts but also on the interaction within different fields of research. Promoting an interaction between different disciplines in natural sciences and enabling young scientists to be involved in it is the general philosophy behind the Biophysical Summer Schools organized by the Rudjer Bošković Institute, Zagreb, Croatia and the Croatian Biophysical Society every 3 years, since 1981.

The International Summer Schools on Biophysics have a broad scope devoted to the structure-function relationship of biological macromolecules and to mayor biophysical techniques. They are internationally recognized and successfully established under the title "Supramolecular Structure and Function" and are included into the curricula of doctoral studies at distinguished European universities. The intention has remained the same through all the ten Schools – to organize courses which provide advanced training at doctoral or postdoctoral level in biosciences. The Schools have gained reputation for running Discussion Clubs as extra curricular activities, where students would invite their peers to gather around lecturers and discuss various topics of specific interest. The enthusiasm of these discussions is always equally shared by students and lecturers. The contributions presented at the Summer School by prominent lecturers illustrate the principles, concepts and methods of biophysics coupled with molecular biology approaches. Given the considerable diversity of topics it covers, we believe that the book will be of interest to scientists involved in different disciplines, as it was to the audience at the Summer School.

The tenth Summer School, as Master Classes of UNESCO, was supported by UNESCO and could be considered as a part of the mosaic forming the European Research Area (ERA) and the European Higher Education Area (EHEA).

The organizers of the International Summer School on Biophysics hope that the publication of this volume and its distribution within the scientific community will serve towards the objectives of expanding, sharing and providing easy access to scientific knowledge in the field of biophysics.

The support to the School by IUPAB and EBSA reflects the international and European interest to bring together scientists of different profiles from all over the world. The national financial supporters were the Ministry of Science, Education and Sport of the Republic of Croatia, the Croatian Academy of Sciences and Arts, The Adris Foundation and The National Foundation for Science, Higher Education and Technological Development of the Republic of Croatia whose substantial support enabled the participation of young scientists from Croatia.

This volume will inform the broader scientific community on the profile of the Summer School and new biophysical achievements, but the most valuable outcome of the tenth School is the exchange of knowledge and friendships established between lecturers and participants in the pleasant atmosphere of the Crveni otok near Rovinj, Croatia.

Zagreb, Croatia

Greta Pifat-Mrzljak
Director of the Schools

The *spiritus movens* of all ten Schools during the period of 30 years was professor Greta Pifat-Mrzljak. Unofortunately, she passed away December 11, 2009. Till the last she was involved in organization and preparation of the School and this proceeding. The tradition, she established, was to present the School with book of selected lectures held on School by distinguished lectures. The intention was to acquaint the brooder scientific communion with the hot subjects in biophysical research. The result of 30 years devotement is the serial of ten books "Supramolecular Structure and Function" as a nice history of biophysical research development and a great help in education of young scientist in that field. Detailed information about Schools and books can be found on web site of the School

http://www.irb.hr/events/confpages/biophysics/

For the outstanding record of accomplishments and leadership of the triennial international summer schools and textbooks on Supramolecular Structure and Function prof. Greta Pifat-Mrzljak was presented the 2010 Emily M. Gray Award by The Biophysical Society.

Zagreb, Croatia

Jasminka Brnjas-Kraljević
on behalf of Organizing Committee

Contents

Contributors

Cesira de Chiara National Institute for Medical Research – MRC, NW7 1AA London, UK, cdechia@imr.mrc.ac.uk

László Forró Laboratory of Physics of Complex Matter, Ecole Polytechnique Fédérale de Lausanne, 1015 Lausanne, Switzerland, Laszlo.forro@epfl.ch

Eric R. Geertsma Department of Biochemistry, Groningen Biomolecular Sciences and Biotechnology Institute, Netherlands Proteomics Centre and Zernike Institute for Advanced Materials, University of Groningen, Nijenborgh 4, 9747 AG Groningen, The Netherlands, e.geertsma@bioc.uzh.ch

Joachim Heberle Experimental Molecular Biophysics, Department of Physics, Free University of Berlin, D-14195 Berlin, Germany, jheberle@zedat.fu-berlin.de

Lenke Horváth Department of Medicine, Unit of Anatomy, University of Fribourg, 1700 Fribourg, Switzerland; Laboratory of Physics of Complex Matter, Ecole Polytechnique Fédérale de Lausanne, 1015 Lausanne, Switzerland, lenke.horvath@epfl.ch

Sergey Ivanchenko Department for Chemistry and Biochemistry, Center for Nanoscience (CeNS) and Munich Center for Integrated Protein Science (CiPSM), Ludwig-Maximilians-Universität München, 81377 Munich, Germany, sergey.ivanchenko@cup.uni-muenchen.de

Don C. Lamb Department for Chemistry, Center for Nanoscience (CeNS) and Munich Center for Integrated Protein Science (CiPSM), Ludwig-Maximilians-Universität München, 81377 Munich, Germany; Department of Physics, University of Illinois at Urbana-Champaign, Urbana, IL 61801, USA, d.lamb@lmu.de

Arnaud Magrez Laboratory of Physics of Complex Matter, Ecole Polytechnique Fédérale de Lausanne, 1015 Lausanne, Switzerland, arnaud.magrez@epfl.ch

Ravi K.R. Marreddy Department of Biochemistry, Groningen Biomolecular Sciences and Biotechnology Institute, Netherlands Proteomics Centre and Zernike

Institute for Advanced Materials, University of Groningen, Nijenborgh 4, 9747 AG Groningen, The Netherlands, r.k.r.marreddy@rug.nl

Kuniaki Nagayama National Institute for Physiological Sciences, Okazaki City, Aichi 444-8787, Japan; The Graduate University for Advanced Studies, School of Physiological Sciences, Hayamacho, Kanagawa 240-0193, Japan, nagayama@nips.ac.jp

Annalisa Pastore National Institute for Medical Research – MRC, NW7 1AA, London, UK, apastor@nimr.mrc.ac.uk

Božidar Pavelić School of Dental Medicine, University of Zagreb, Zagreb, Croatia, bozidar.pavelic@zg.hinet.hr

Ognjen Pavičević Biozon doo, Buzinski prilaz 10 10 000 Zagreb, Croatia, pavicevic@biozon.hr

Davor Pavuna Physics Section, EPFL, CH-1015 Lausanne, Switzerland, davor.pavuna@epfl.ch

Bert Poolman Department of Biochemistry, Groningen Biomolecular Sciences and Biotechnology Institute, Netherlands Proteomics Centre and Zernike Institute for Advanced Materials, University of Groningen, Nijenborgh 4, 9747 AG Groningen, The Netherlands, b.poolman@rug.nl

Domagoj Prebeg Biozon doo, Buzinski prilaz 10 10 000 Zagreb, Croatia, domagoj.prebeg@ozoneacademy.org

Ionela Radu Experimental Molecular Biophysics, Department of Physics, Free University of Berlin, D-14195 Berlin, Germany, iradu@zedat.fu-berlin.de

Michael Schleeger Experimental Molecular Biophysics, Department of Physics, Free University of Berlin, D-14195 Berlin, Germany, michael.schleeger@fu-berlin.de

Beat Schwaller Department of Medicine, Unit of Anatomy, University of Fribourg, 1700 Fribourg, Switzerland, beat.schwaller@unifr.ch

Ian C.P. Smith Institute for Biodiagnostics, National Research Council Winnipeg, Winnipeg, MB, Canada R3B 1Y6, ian.smith@nrc-cnrc.gc.ca

Ray L. Somorjai Institute for Biodiagnostics, National Research Council Winnipeg, Winnipeg, MB Canada, R3B 1Y6, ray.somorjai@nrc-cnrc.gc.ca

Piero Andrea Temussi Dipartimento di Chimica, Universita' di Napoli Federico II, I-80126 Napoli, Italy; National Institute for Medical Research – MRC, NW7 1AA London, UK, temussi@unina.it

Mario Zovak Croatia School of Medicine, University of Zagreb, Zagreb, Croatia, zovak@biozon.hr

Fluorescence Correlation Spectroscopy: Principles and Developments

Sergey Ivanchenko and Don C. Lamb

Abstract Twenty years ago, fluorescence measurements at low concentrations were difficult due to the weak fluorescence signal and intrinsic fluctuations of the sample. With the development of FCS and its implementation on a confocal microscope, it is possible to use the inherent fluctuations to gain information over the concentration, molecular brightness, microscopic rate constants for reactions and mobility of the measured sample. In recent years, there has been a strong increase in the development and application of fluctuation methods. With pulsed interleaved excitation, stoichiometry information can be obtained and spectral cross-talk can be eliminated from FCCS experiments. An elegant implementation of two-focus FCS has also been introduced to allow absolute measurements of diffusion coefficient without precise knowledge of the psf of the microscope and is less sensitive to the laser excitation intensity and saturation effects. Scanning methods such as Scanning FCS and RICS increase the effective volume, which is advantageous for live-cell measurements where diffusion is slow and photobleaching is a problem. In this article, describe the basics of FCS and its limitations as well as a short discussion of a handful of emerging techniques. There are still many other equally interesting applications of fluorescence fluctuation spectroscopy that we have not been able to touch upon. And, if the past is any indication of the future, there will be a number of novel fluorescence fluctuation spectroscopy methods emerging in the near future.

Keywords Fluorescence correlation spectroscopy (FCS) · ACF · ALEX · ccRISC · FRET · PIE

D.C. Lamb (✉)
Department for Chemistry, Center for Nanoscience (CeNS) and Munich Center for Integrated Protein Science (CiPSM), Ludwig-Maximilians-Universität München, 81377 Munich, Germany; Department of Physics, University of Illinois at Urbana-Champaign, Urbana, IL 61801, USA
e-mail: d.lamb@lmu.de

J. Brnjas-Kraljević, G. Pifat-Mrzljak (eds.), *Supramolecular Structure and Function 10*, DOI 10.1007/978-94-007-0893-8_1,

Abbreviations

2fFCS	Two-focus fluorescence correlation spectroscopy
ACF	Autocorrelation function
ALEX	Alternating laser excitation
CCF	Cross-correlation function
ccRISC	Cross-correlation raster image correlation spectroscopy
FCS	Fluorescence correlation spectroscopy
FCCS	Fluorescence cross-correlation spectroscopy
FRET	Förster resonance energy transfer
PIE	Pulsed interleaved excitation
RISC	Raster image correlation spectroscopy

1 Introduction

Approximately 100 years ago, the first fluorescence microscopes were built (Heimstadt 1911, Reichert 1911, Lehman 1913). Fluorescence has many advantages for investigating biological systems; for example, cells are typically transparent to visible light and fluorescence experiments can be performed without direct contact with the sample. Often, background fluorescence is low and fluorescence microscopy can be performed with high contrast. In addition, the fluorescence signal can be detected with high sensitivity, as single-photon counting detectors with high quantum yield are currently available. It is possible to perform fluorescence experiments over a broad range of concentrations that extends down to the single molecule scale due to advances in detector sensitivity, aberration free optics and the development of stable light sources. When performing experiments with molecules in solution at low concentrations (e.g. in a cuvette or in the focus of a confocal microscope), the detected fluorescence signal is noisy. This noise does not depend on the quality of the detectors or the stability of the excitation sources, but arises from fluctuations in the number of fluorescent molecules in the observation volume. Due to laws of thermodynamics, the number of molecules in the detection volume constantly fluctuates, giving rise to fluctuations in the detected fluorescence signal. Thermodynamic fluctuations were first observed experimentally with gold beads already in 1911 (Svedberg and Inouye 1911), verifying the predications of Einstein (Einstein 1905) and von Smoluchowski (von Smoluchowski 1906). Fluctuations contain interesting dynamic information regarding the sample and can be extracted from the data with the appropriate methods. To this end, a correlation approach has been developed. Events that are correlated, such as the detection of multiple photons from the same molecule traversing the observation volume, will show up in a correlation analysis. Thereby, information regarding the mobility and average number of the fluorescent molecules in the observation volume can be determined.

Fluorescence Correlation Spectroscopy (FCS) was first performed by Madge et al. (1972). In the first years, the group published three seminal works on FCS including the theory for freely diffusing particles, the unimolecular and bimolecular reactions (Elson and Magde 1974), experimental realization of the method

(Magde et al. 1974) and the expansion of FCS for systems under flow (Magde et al. 1978). In the early days of FCS, Ehrenberg and Rigler expanded the theory of FCS to describe rotational Brownian motion (Ehrenberg and Rigler 1974, 1976). A thorough description of the error analysis involved in FCS was published by Koppel (1974). Initially, FCS required long measurement times and had a low signal-to-noise ratio due to the low detection efficiencies, high background and large volumes used in the initial systems. A significant improvement came when FCS was combined with confocal microscopy that was first implemented by Koppel and coworkers (1976) and later championed by Rigler and coworkers in the 1990s (Rigler et al. 1993, Eigen and Rigler 1994, Widengren et al. 1994, 1995).

Today, FCS is widely applied in a broad number of disciplines including physics, chemistry, biology, biophysics, biochemistry and medicine. The apparatus is commercially available and an FCS signal is easy to obtain. Anything that alters the fluorescence intensity in a correlated way will show up in an FCS measurement. This is one of the advantages of FCS but is also an aspect that requires caution. This is especially true for measurements of slowly diffusion particles or for FCS measurements in living cells where many external variables such as mechanical vibrations or oscillations in laser intensity can contribute to the correlation function.

2 Principles of FCS

2.1 *What Is FCS?*

FCS has been used to measure a large number of phenomena including translational diffusion (Magde et al. 1974), rotational diffusion (Ehrenberg and Rigler 1974, Aragón and Pecora 1976, Kask et al. 1989), triplet-state dynamics (Widengren et al. 1994, 1995), chemical reactions (Magde et al. 1974, Magde 1976, Rauer et al. 1996, Lamb et al. 2000a, Bismuto et al. 2001) and conformational fluctuations (Bonnet et al. 1998, Torres and Levitus 2007). The fundamental process upon which FCS is based is the non-stochastic nature of the fluorescence photons being detected. For example, in translational diffusion, an increase in the number of detected photons is observed when a fluorescent molecule enters the observation volume and again the signal decreases when the molecule diffuses away. This correlation is extracted using a correlation analysis. The typical FCS experiment is based upon two assumptions: (1) the system is stationary, meaning that the average values of the phenomena being measured do not change with time and (2) the system is ergodic. Hence, every sizable sampling of the process is representative of the whole. The details of what FCS is and how it works are discussed below.

2.1.1 Fluorescence

FCS is based upon fluorescence. Fluorescence is the property of a molecule to emit light upon returning to the ground state from the lowest level of the singlet excited state after optical excitation (Fig. 1a). Molecules that can emit light upon such an

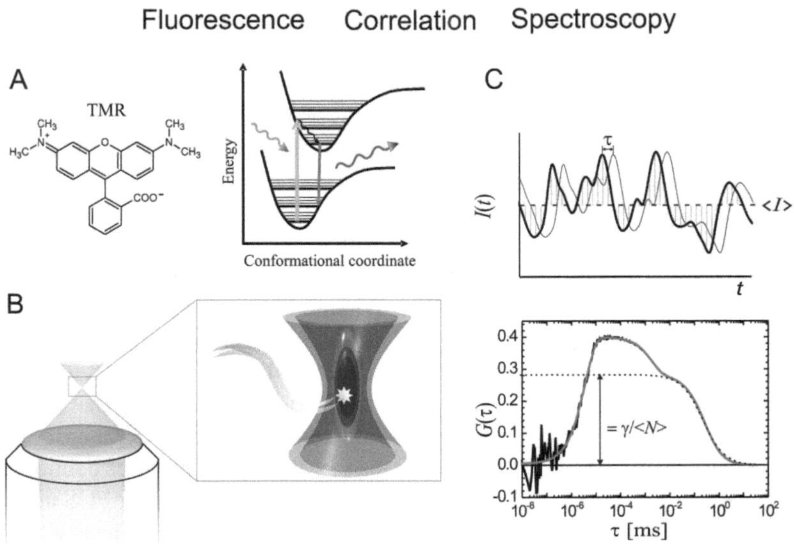

Fig. 1 Fluorescence Correlation Spectroscopy. FCS is based upon fluorescence. **A** The chemical structure of a typical fluorescent molecule, tetramethylrhodamine, along with the Franck-Condon diagram of the electronic transition. **B** A schematic diagram of the focus of a confocal microscope with a blow up of the excitation (*cone shaped*) and detection (*ellipsoidal*) volumes. A molecule diffusing through the confocal volume is shown as a *star*. **C** The self-similarity of the fluorescence time series (*upper panel*) is reflected in its autocorrelation function (*lower panel*). ACF from a diffusing molecule shows several processes that occur during its passage through the confocal volume; anti-bunching, triplet-state excitation and translational diffusion. The amplitude of the translation diffusion component is inversely proportional to the total number of fluorescent molecules in the confocal volume. The plot is adopted from (Felekyan et al. 2005)

electronic transition are called fluorophores (e.g. tetramethylrhodamine, Fig. 1a). The fluorophores that are typically used contain an extended π-electron conjugated system in which electrons can move freely. Such conjugated π-electron systems have a large cross section for absorption of visible light and upon absorption of a photon, an electron is transferred into an electronic excited state. The electron relaxes back to the ground state within nanoseconds, giving up the absorbed energy in a form of either a photon or a phonon. Different fluorophores have different fluorescent properties such as different excitation and emission spectra, and these differences can be exploited to investigate multiple species and their interactions simultaneously.

Typically, FCS is performed on a fluorescence confocal microscope[1]. See e.g. (Webb 1996) for a review on confocal microscopy. The focal size of the confocal microscope is limited via diffraction to roughly 1 fL. The optical response of the microscope to a point source at the center of the focus is referred to as the

[1] It is also possible to perform FCS using Total Internal Reflection Excitation, but a description of this method is beyond the scope of this publication. For details see Thompson et al. (1981).

point-spread-function (psf) or observation volume and is approximated by a three-dimensional Gaussian with different lateral and axial dimensions (Fig. 1b). The observation volume is the overlap between the excitation, sample and detection volumes and is given by:

$$W(\mathbf{r}) = I_0(0)\exp\left[-\frac{2\left(x^2+y^2\right)}{w_r^2} - \frac{2z^2}{w_z^2}\right] \tag{1}$$

where w_r and w_z are the radial and axial distance from the center of the psf to where the intensity has decreased by $1/e^2$. If we assume that we have freely diffusing, non-interacting particles that do not undergo photophysical effects, we can determine the total measured fluorescence intensity by the position of the particles as a function of time. The fluorescence signal is then given by:

$$F(t) = \kappa\sigma\phi \int d\mathbf{r}\ W(\mathbf{r})C(\mathbf{r},t), \tag{2}$$

where κ is the overall detection efficiency of the system, σ is the absorption cross section at the wavelength of excitation, ϕ is the fluorescence quantum yield of the fluorophore and $C(\mathbf{r},t)$ represents the concentration of particles at position $\mathbf{r}$ and time t. The product:

$$\varepsilon = \kappa\sigma\phi W\,(\mathbf{0}), \tag{3}$$

yields the molecular brightness of the fluorophore at the center of the psf.

Other methods exist that utilize the information obtainable from equilibrium fluctuations of the sample that do not rely on fluorescence. Dynamic light scattering, for example, detects photons scattered from a sample, which are correlated to determine the translational diffusion coefficient of molecules. However, FCS is more sensitive and can be performed at lower concentrations than dynamic light scattering.

2.1.2 Correlation

The heart of FCS is the correlation analysis. It is the correlation function that allows us to extract information regarding the fluctuations. The temporal autocorrelation function (ACF), also referred to as the normalized second-order autocorrelation function, is defined as:

$$G(\tau) = \frac{\langle F(t)F(t+\tau)\rangle - \langle F(t)\rangle^2}{\langle F(t)\rangle^2} = \frac{\langle \delta F(t)\delta F(t+\tau)\rangle}{\langle F(t)\rangle^2} \tag{4}$$

where $\langle\rangle$ refers to the time averaged value and $\delta F(t) = F(t) - \langle F(t)\rangle$. The denominator renormalizes the ACF to the average fluorescence intensity. The ACF measures the self-similarity of the fluorescence intensity time series as a function of the delay time τ, also referred to as the correlation time (Fig. 1c). At zero delay, the amplitude of the ACF is given by:

$$G(0) = \frac{\langle \delta F(0) \delta F(0) \rangle}{\langle F \rangle^2} = \frac{\sum_{i=1}^{\ell} (F(t_i) - \langle F \rangle)^2 \Big/ \ell}{\left(\sum_{i=1}^{\ell} F(t_i) \Big/ \ell \right)^2} = \frac{\sigma^2}{\mu^2} \tag{5}$$

where σ^2 is the variance of the time series and μ the average fluorescence intensity. Due to the definition of the ACF (Eq. 4), it has a maximum value at $G(0)$. This can be seen from the fact that when there is no shift, (i.e. $\tau = 0$), the maximum value of $\delta F(t)$ will be multiplied by itself summed with the second maximum value of $\delta F(t)$ multiplied by itself and so on. All points will add constructively as $\delta F(t)^2 > 0$. For non-conserved, non-periodic signals, $G(\tau) \to 0$ as $\tau \to \infty$.

Assuming a 3D Gaussian observation volume for $W(\mathbf{r})$ (Eq. 1), using Eq. (2) for the fluorescence intensity, the ACF (Eq. 4) can be solved analytically and is given by:

$$G_D(N, D, \tau) = \frac{\gamma}{\langle N \rangle} \left(\frac{1}{1 + \tau/\tau_D} \right) \left(\frac{1}{1 + (w_r/w_z)^2 \tau/\tau_D} \right)^{1/2} \tag{6}$$

$$\text{where} \quad \tau_\text{D} = \frac{w_r^2}{4D} \text{ or } \tau_\text{D} = \frac{w_r^2}{8D} \tag{7}$$

for one- and two-photon excitation respectively and γ is a factor that depends on the geometry of the observation volume ($\gamma = 2^{-3/2}$ for a 3D Gaussian). See Section 2.2 for more details regarding the γ factor.

In literature, a second definition of the ACF is also used:

$$g(\tau) = \frac{\langle F(t) F(t + \tau) \rangle}{\langle F(t) \rangle^2}. \tag{8}$$

For a 3D Gaussian observation volume, the ACF in this form is given by:

$$g(\tau) = 1 + \frac{\gamma}{\langle N \rangle} \left(\frac{1}{1 + \tau/\tau_D} \right) \left(\frac{1}{1 + (w_r/w_z)^2 \tau/\tau_D} \right)^{1/2}. \tag{9}$$

In this representation, the ACF is proportional to the probability of detecting a photon at the time τ, given that a photon was detected at $\tau = 0$. At long times, when no correlation is observable, the probability of detecting a photon is constant and equal to the random possibility of a second photon being detected depending on the average count rate.

2.1.3 Spectroscopy

The remaining term in the name of the method is spectroscopy. In particular, we are performing spectroscopy on the fluctuations and relate the properties of the fluctuations to properties of the fluorescent molecules. The ACF for Rhodamine 110 freely diffusing in solution is shown at the bottom of Fig. 1c. Several processes influence the shape of the curve. At shorter times ($\sim 10^{-9}$ s), the lack of correlation is caused by a non-zero delay between the absorption and emission of a photon. During this delay, no further absorption-emission events are possible, which leads to loss of correlation. This rising edge of the ACF is referred to as the antibunching term and is typically not recorded in most experiments. However, the early times are useful for precise measurements of polarization anisotropies of fluorophores.

The maximum of the ACF (Fig. 1c) occurs around 10^{-5} ms and decays in two characteristic steps. The microsecond decay is typically related to relaxation of the fluorophore from the triplet state and its amplitude depends on excitation power. The amplitude increases with increasing power. For larger molecules such as proteins, other processes like rotation also contribute to the ACF on this timescale. Measurements at such short times are difficult due to distortions introduced by the detector (e.g. detector dead time or detector afterpulsing). To measure the early correlation times, two detectors are used where the light is split equally between the two detectors and cross-correlation of the signal from the two detectors is performed (Brown and Twiss 1956).

The second step in the ACF seen at longer times is caused by translational diffusion and depends on the average time the molecule spends in the observation volume. The average duration of the fluctuations, given by the decay time of the ACF (τ_D), can be related to the translation diffusion coefficient of the fluorescent molecule. From Eq. (7), D can be calculated if ${w_r}^2$ is known or, if the diffusion coefficient is known (for example, $D = 414 \pm 5\ \mu\text{m}^2/\text{s}$ for Rhodamine 6G in buffer at 25°C (Muller et al. 2008)), the size of the psf can be determined. As D and ${w_r}^2$ always appears together in Eq. (7), D can only be determined to the accuracy to which ${w_r}^2$ is known. This difficulty can be overcome by bringing an absolute distance into the equation as is the case of two-focus FCS described in Section 3. Fluctuations other than translation diffusion, which occur on different time scales, can also be analyzed with FCS and related to physical parameters such as the triplet-state-lifetime or the microscopic rate coefficients.

The amplitude of the fluctuations also provides relevant information regarding the fluorescent molecules such as the fraction of molecules in the triplet state or information regarding the equilibrium coefficient of a unimolecular reaction. As the ACF is renormalized to the intensity, the amplitude of the fluctuations due to translational diffusion and thus also the corresponding ACF, is inversely proportional to the average number of particles in the observation volume, N (Fig. 1c, bottom panel). To compare the number of particles with the actual concentration in mol/L, the geometric factor γ needs to be considered.

2.2 The Geometrical Factor γ

The difficulty in converting the number of particles in the observation volume into a concentration is determining the size of the observation volume. In the ideal case, the fluorescence intensity emitted from fluorophores inside the observation volume is constant and zero when the molecule is outside the observation volume. Thus, the volume of the psf is well defined and $\gamma = 1$. However, when the fluorescence intensity decays within the psf, the definition of a volume becomes arbitrary (Fig. 2a). To illustrate this, consider the one-dimensional case where we need to define a measure of the length, w_x, of a Gaussian function. The tails of the Gaussian become zero only when x reaches infinity. To estimate a length, typically a box function is assumed (Fig. 2b). The area under the curve, which is proportional to the total fluorescence intensity, is chosen to be the same as for the Gaussian function. However, different amplitudes for the box function can be chosen, and hence different length scales result. For the observation volume in FCS, the same situation applies in three dimensions.

Many research groups define the amplitude of the correlation function to be equal to the inverse number of molecules, that is $G(0) = 1/N$ and $\gamma = 1$. With this approach, the effective volume is given by:

$$V_{eff} = \pi^{\frac{3}{2}} w_r^2 w_z. \tag{10}$$

A second approach is to define a volume of uniform illumination where the fluorescence intensity within the volume is equal to the peak value of the Gaussian psf. The molecular brightness for a molecule at the center of the psf can be defined as in Eq. (3) and the total average intensity is then given by:

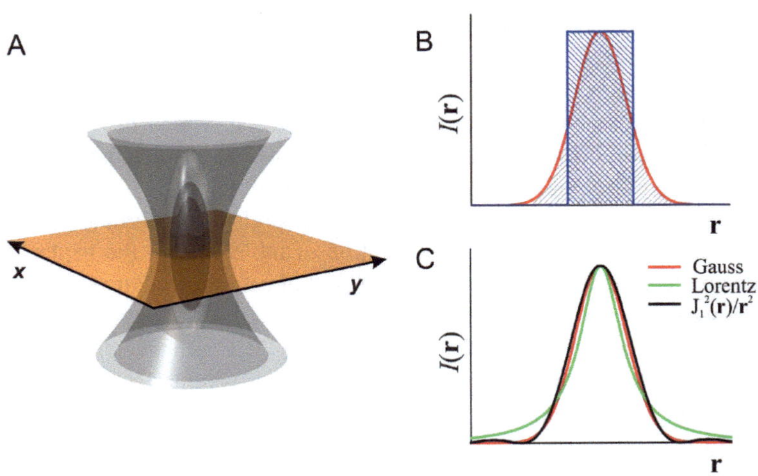

Fig. 2 The point spread function and the confocal volume. **A** A three-dimensional representation of the confocal observation volume. **B** A one-dimensional slice through the lateral dimension of the three-dimensional Gaussian used to approximate the psf (*red*). The top-hat function used to approximate the effective one-dimensional confocal volume is shown in *blue*. **C** The one-dimensional intensity distribution profiles a Gaussian (*red*), Lorenzian (*green*) and $\sim J_1{}^2(\mathbf{r})/\mathbf{r}^2$ (*black*), where $J_1(\mathbf{r})$ is the first order Bessel function, which are typically used to describe the psf of a confocal volume

$$\langle F \rangle = \varepsilon \langle N \rangle . \tag{11}$$

The effective 3D Gaussian volume is given by:

$$V_{eff} = \left(\frac{\pi}{2}\right)^{\frac{3}{2}} w_r^2 w_z, \tag{12}$$

and the geometrical factor is given by $\gamma = 2^{-\frac{3}{2}}$ or 0.35355 or, in general, by:

$$\gamma = \frac{\int d\mathbf{r}(W(\mathbf{r})/W(0))^2}{\left[\int d\mathbf{r}\,(W(\mathbf{r})/W(0))\right]} \tag{13}$$

For the 2D Gaussian-Lorentzian approximation used with two-photon excitation, the geometrical factor γ is 0.185. The sharper the boundary of the psf is, the larger the fluctuations will be and hence, the higher the amplitude of the ACF will be for the same sample concentration. Therefore, the Gaussian function, which has a steeper decay compared to Lorentzian (Fig. 2c), has a higher geometrical factor. Technically, a Gaussian profile works well when exciting with a Gaussian beam that under fills the back aperture of the objective. When the excitation beam overfills the back aperture of the objective, the transversal shape of the psf becomes $\sim J_1{}^2(\mathbf{r})/\mathbf{r}^2$, where $J_1(\mathbf{r})$ is the first order Bessel function, shown in Fig. 2c for comparison.

As the observation volume is defined via the choice of the geometrical factor, various definitions are taken by different groups. Depending on the definition chosen for γ, the number of molecules in the observation volume will vary as well as the definition of the effective volume and the meaning of the molecular brightness. These differences need to be correctly accounted for when calculating absolute concentration values. However, the molecular brightness and number of molecules are related depending on the definition of the volume. Hence, one can achieve consistent results between an FCS and, for example, the Photon Counting Histogram (PCH) analysis when a consistent γ-factor is used.

2.3 The Signal-to-Noise in FCS

As the fluctuations in fluorescent intensity are what are measured in FCS, FCS measurements behave differently than other fluorescence measurements. In general, the signal-to-noise in an FCS experiment is an involved calculation (see Koppel 1974). In the case where the total number of photons in not limiting (the high intensity limit), the uncertainty of the experiments is dominated by the limited number of fluctuations measured. The signal-to-noise can be approximated by:

$$\frac{S}{N} \approx \left(\frac{t_{\exp}}{\tau_C}\right)^{\frac{1}{2}} \tag{14}$$

where t_{exp} is the experimental measurement time and τ_C is the correlation time of the fluctuations. In this case, the experiment can only be improved by longer measurement times.

In typical FCS experiments, the number of detected photons is the limiting factor (at least for the early time points). The signal-to-noise ratio will then depend on the experimental measurement time, the average fluorescence intensity and the amplitude of the autocorrelation function:

$$\frac{S}{N} \approx \left(t_{\text{exp}}\right)^{\frac{1}{2}} \langle F \rangle \frac{\gamma}{\langle N \rangle} \tag{15}$$

The fluorescence intensity is given by $\langle F \rangle = \varepsilon \langle N \rangle$, thus, the signal-to-noise ratio is given by:

$$\frac{S}{N} \approx \left(t_{\text{exp}}\right)^{\frac{1}{2}} \varepsilon\, \gamma \tag{16}$$

To improve the signal-to-noise ratio, one can either extend the measurement time, increase the molecular brightness or change the geometry. However, the signal-to-noise of FCS is independent of sample concentration (at least over a wide range of concentrations where other affects can be ignored).

2.4 The Limitations of FCS

Although FCS is independent of concentration, there are some practical limitations. The lower measurable concentrations are on the order of a few pM. It is possible to measure FCS in samples with concentrations in the low pM range by using scanning FCS to increase the effective volume (as will be discussed in Sections 3.1 and 3.3) or with a lot of patience. The experimental measurement times become quite long because the number of fluctuations measured is limited. In addition, correlation coming from background in the buffer becomes significant (see Section 2.6 on FCS with multiple species). The upper concentration limit is typically on the order of a couple hundred nanomolar. For samples with concentrations higher than ~50 nM, the APD detectors used for FCS begin to saturate. Hence, low laser excitation intensities need to be used and the molecular brightness of the measured fluorophores drops. Also the amplitude of the ACF is very low and the contribution from other correlated signals becomes significant. When taking these practical limitations into account, the optimal concentration for FCS experiments is typically between 1 and 10 nM.

The timescales of the fluctuations that can be measured with FCS range from ps to s. Under normal conditions, the earliest timescales that can be measured with FCS are 10–100 ns limited by the number of detected photons. When afterpulsing of the detectors is severe or the fluorescence intensity of the sample is low, afterpulsing can be observed into the microsecond time regime of the ACF (Bismuto et al.

2001). Using a Hanbury-Brown–Twiss setup (Brown and Twiss 1956), the deleterious effects of detector dead time and afterpulsing can be avoided. When using high excitation intensities (which typically distort the translational diffusion portion of the FCS curve), dynamics on the ps to ns timescale such as photon anti-bunching or rotation can be measured with FCS (Felekyan et al. 2005). The maximum timescales of the dynamics depend on how long the molecule remains in the observation volume. Small proteins in a well-focused confocal volume diffuse through on the ms timescale. By suboptimal focusing of the laser through the objective, the diffusion time can be extended to $\sim$ 10 ms. When the molecules are embedded in viscous solvents, gels or trapping in vesicles (Dickson et al. 1997, Lu et al. 1998, Boukobza et al. 2001, Rhoades et al. 2003), the diffusion time can be extended into the second regime. Longer timescales can be obtained with FCS by immobilizing the molecules (Wennmalm et al. 1997, Ha et al. 2002).

2.5 Reactions

2.5.1 Triplet-State

Any process that leads to non-random variations in fluorescence intensity can be investigated using FCS. This includes the triplet-state photophysics of fluorophores. When molecules go into the triplet state, they become trapped and cannot fluoresce again until they first return to the ground state, which occurs on the order of μs. Hence, the fluorescence intensity drops. The dynamics of the triplet-state in FCS was described by Widengren (Widengren et al. 1994, 1995). Using a simple three-state model for the excitation scheme:

$$\frac{\partial}{\partial t}\begin{pmatrix} S_0(\mathbf{r},t) \\ S_1(\mathbf{r},t) \\ T_1(\mathbf{r},t) \end{pmatrix} = \begin{pmatrix} -k_{12}(\mathbf{r},t) & k_{21} & k_{31} \\ k_{12}(\mathbf{r},t) & -(k_{12}+k_{23}) & 0 \\ 0 & k_{23} & -k_{31} \end{pmatrix} \begin{pmatrix} S_0(\mathbf{r},t) \\ S_1(\mathbf{r},t) \\ T_1(\mathbf{r},t) \end{pmatrix} \tag{17}$$

where S_0 , S_1 and T_1 are the populations of the ground state, singlet excited state and triplet state respectively, $k_{12}(\mathbf{r},t) = \sigma W(\mathbf{r},t)$ is the excitation rate, k_{21} and k_{31} are the rates at which molecules in the excited singlet and triplet state return to the ground state and k_{23} is the rate of intersystem cross from the singlet to the triplet state. Assuming that the diffusion time is well separated from the triplet timescale and that the fluorescence rate is much faster than the rate of intersystem crossing and the triplet state lifetime ($k_{21} >> k_{23},\ k_{31}$), the ACF can be calculated (assuming a 3D Gaussian observation volume):

$$G(\tau) = \frac{\gamma}{\langle N\rangle}\left(\frac{1}{1+\tau/\tau_D}\right)\left(\frac{1}{1+\left(w_r/w_z\right)^2\tau/\tau_D}\right)\left(1+\frac{T_{eq}}{1-T_{eq}}e^{-\lambda_T t}\right), \tag{18}$$

where T_{eq} is the fraction of molecules in the triplet state in equilibrium, λ_T reflects the timescale of the triplet state relaxation and is given by:

$$\lambda_T = \left(k_{31} + \frac{k_{12}(\mathbf{r},t)k_{23}}{k_{12}(\mathbf{r},t) + k_{21}}\right) \approx k_{31}. \tag{19}$$

2.5.2 Unimolecular Reactions

Many biomolecules undergo conformational motions that can lead to quenching of a fluorophore. The conformational flucutations can be described by a unimolecular reaction model,

$$A \underset{k_b}{\overset{k_f}{\Leftrightarrow}} B.$$

As the fluorescence intensity changes upon quenching, such fluctuations will be observable in the ACF. A two-state model describing the unimolecular reaction is given by:

$$\frac{\partial}{\partial t}\begin{pmatrix} C_A(\mathbf{r},t) \\ C_B(\mathbf{r},t) \end{pmatrix} = \begin{pmatrix} D_A\nabla^2 - k_f & k_b \\ k_f & D_B\nabla^2 - k_b \end{pmatrix}\begin{pmatrix} C_A(\mathbf{r},t) \\ C_B(\mathbf{r},t) \end{pmatrix} \tag{20}$$

where k_f and k_b are the forward and backward rate constants, and D_A and D_B are the diffusion coefficients of the complex in the A and B conformation. Assuming that the diffusion behaviour of the complex is the same in both conformations, the ACF for a diffusion species undergoing a unimolecular reaction is given by (Elson and Magde 1974, Thompson 1991, Lamb et al. 2000b):

$$G(\tau) = G_D(\tau, N_A + N_B, \tau_D)\left[1 + K\left(\Im_A - \frac{\Im_B}{K}\right)e^{-\lambda\tau}\right], \tag{21}$$

where $K = k_f/k_b$ is the equilibrium coefficient, $\lambda = k_f + k_b$ is the apparent reaction rate coefficient and $\Im_i$ is the fractional intenstiy of the ith species:

$$\Im_i = \frac{\varepsilon_i \langle N_i \rangle}{\varepsilon_A \langle N_A \rangle + \varepsilon_B \langle N_B \rangle}. \tag{22}$$

2.5.3 Bimolecular Reactions

Binding of a fluorophore to another biomolecule can lead to a large enhancement in fluorescence intensity as is observed for the binding of ethidium bromide to DNA or 1-anilinonaphthalene-8-sulfonic acid (ANS) binding to hydrophobic pockets of proteins. Hence, such reactions can also be measured using correlation spectroscopy. The chemical equation of a bimolecular reaction is given by:

$$M + L \underset{k_{off}}{\overset{k_{on}}{\Leftrightarrow}} ML \tag{23}$$

where M represents a macromolecule, L the ligand, and ML the complex of macromolecule and ligand. Now we have three species diffusing in an open volume that can interconvert. The concentration of the various species is given by three coupled, differential equations:

$$\frac{\partial}{\partial t}\begin{pmatrix} C_M(\mathbf{r},t) \\ C_L(\mathbf{r},t) \\ C_{ML}(\mathbf{r},t) \end{pmatrix} = \begin{pmatrix} D_M\nabla^2 - k_{on}\langle C_L\rangle & -k_{on}\langle C_M\rangle & k_{off} \\ -k_{on}\langle C_L\rangle & D_L\nabla^2 - k_{on}\langle C_M\rangle & k_{off} \\ k_{on}\langle C_L\rangle & k_{on}\langle C_M\rangle & D_{ML}\nabla^2 - k_{off} \end{pmatrix}\begin{pmatrix} C_M(\mathbf{r},t) \\ C_L(\mathbf{r},t) \\ C_{ML}(\mathbf{r},t) \end{pmatrix}. \tag{24}$$

Assuming that the diffusion coefficient of the macromolecule does not change significantly upon binding of the ligand and that the reaction rate, $k_r = k_{on}(\langle C_M\rangle + \langle C_L\rangle) + k_{off}$, is much faster than the diffusion time of the ligand through the volume, $k_r >> D_L/w_r^2$, the ACF can be determined in a simplified closed form (Lamb et al. 2000b, Bismuto et al. 2001):

$$\begin{aligned} G(\tau) &= G_D(N_M + N_{ML}, D, \tau)(\Im_M + \Im_{ML})^2 \\ &\quad + \frac{\langle C_{ML}\rangle k_{on}}{k_r} G_D\left(N_M + N_{ML}, D^{(+)}, \tau\right) \\ &\quad \left[\Im_M - \frac{\langle C_M\rangle}{\langle C_{ML}\rangle}\Im_{ML} - \frac{\langle C_M\rangle + \langle C_{ML}\rangle}{\langle C_{ML}\rangle}\Im_L\right]^2 \\ &\quad + \frac{\langle C_{ML}\rangle\left(k_{on}\langle C_B\rangle + k_{off}\right)}{\langle C_M\rangle k_r} G_D\left(N_M + N_{ML}, D^{(-)}, \tau\right) \\ &\quad \left[\Im_M - \frac{\langle C_M\rangle}{\langle C_{ML}\rangle}\Im_{ML} + \frac{\langle C_M\rangle + \langle C_{ML}\rangle}{\langle C_{ML}\rangle}\Im_L\right]^2 e^{-k_r\tau} \end{aligned} \tag{25}$$

$$\begin{aligned} \text{where } G_{Di}(N_i, D_i, \tau) &= \frac{\gamma}{\langle N_i\rangle}\left(\frac{1}{1+\tau/\tau_{D_i}}\right)\left(\frac{1}{1+(w_r/w_z)^2\tau/\tau_{D_i}}\right)^{1/2} \\ D^{(+)} &= \left[D_L\frac{\langle C_L\rangle}{\langle C_L\rangle + \langle C_{ML}\rangle f} + D_M\frac{\langle C_{ML}\rangle}{\langle C_L\rangle + \langle C_{ML}\rangle f}\right], \\ D^{(-)} &= \left[D_L\frac{\langle C_{ML}\rangle}{\langle C_L\rangle + \langle C_{ML}\rangle f} + D_M\frac{\langle C_L\rangle}{\langle C_L\rangle + \langle C_{ML}\rangle f}\right], \\ \text{and} \quad f &= \frac{\langle C_M\rangle}{\langle C_M\rangle + \langle C_{ML}\rangle}. \end{aligned} \tag{26}$$

For measurements where the ligand is in large excess, the ACF can be further simplified. The on-rate can then be treated as a pseudo first-order rate constant and the ACF is given by

$$\begin{aligned} G(\tau) = G_D(\tau, N_M + N_{ML}, \tau_D)&\left[(\Im_M + \Im_{ML})^2 + K\langle C_L\rangle\left(\Im_M - \frac{\Im_{ML}}{K\langle C_L\rangle}\right)^2 e^{-k_r\tau}\right] \\ &+ G_D\left(\tau, N_L, \tau_{D_L}\right)\Im_L^2, \end{aligned} \tag{27}$$

where $K = k_{on}/k_{off}$ is the equilibrium coefficient. It is no longer necessary to assume, as in Eq. (25), that the reaction rate be much faster than the diffusion of the ligand through the confocal volume although the diffusion coefficient of the macromolecule should not change significantly in the presence of the ligand. When the ligand does not fluoresce free in solution but only when bound to the macromolecule, Eq. (27) simplifies even further to:

$$G(\tau) = G_D(\tau, N_M + N_{ML}, \tau_D)\left[1 + \frac{1}{K\langle C_L\rangle}e^{-k_r\tau}\right]. \tag{28}$$

This equation is analogous to Eq.(21) where state A is dark, as expected for a pseudo first-order rate equation.

2.6 Multiple Species

FCS works well when a single diffusing species is present in the measurement and when the timescales of any other correlations are well separated. However, when two species are present with similar diffusion coefficients, it becomes difficult to separate them. When the different species have different molecular brightnesses, than a quantitative FCS analysis becomes tricky. This is the case, for example, when a protein exists in an equilibrium between the monomeric and dimeric states.

When measuring FCS with multiple species, the relative brightness of the species must be known to perform a quantitative analysis. The ACF from multiple non-interacting species is given by a linear combination of the ACFs for the individual species, but weighted by the square of the fractional intensities of the individual species:

$$G_{meas}(t) = \sum_{\substack{j \\ \text{over all species}}} \Im_j^2 G_j(\tau) \tag{29}$$

where

$$\Im_i = \frac{\varepsilon_i \langle N_i\rangle}{\sum\limits_{\substack{j \\ \text{over all species}}} \varepsilon_j \langle N_j\rangle}. \tag{30}$$

In Fig. 3, two types of diffusing molecules are observable: a bright, slowly diffusing species and a faster diffusing species with 1/4th the brightness. Assuming that each species exist at the same concentration, the ACFs for the individual species would have the same amplitude. However, when both species are present simultaneously, the situation becomes more complicated. If the two species had the sample molecular brightness, the amplitude would decrease by factor 2, since the total number of molecules in the volume has doubled. For the case shown in Fig. 3, the

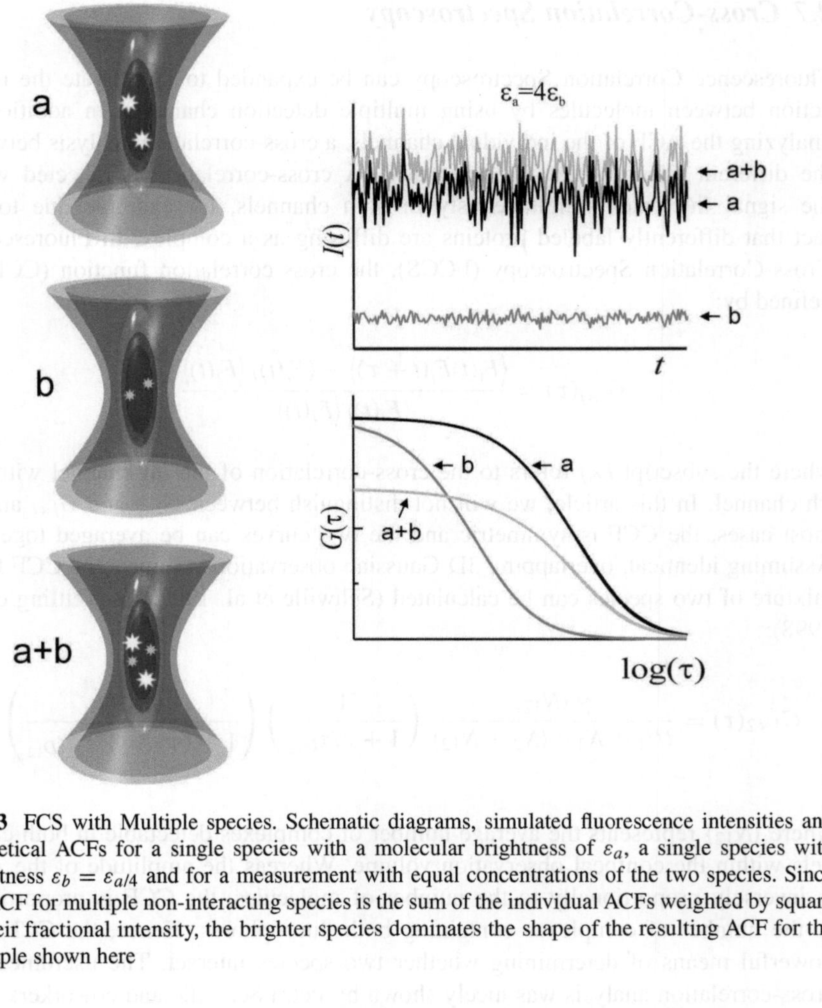

Fig. 3 FCS with Multiple species. Schematic diagrams, simulated fluorescence intensities and theoretical ACFs for a single species with a molecular brightness of ε_a, a single species with brightness $\varepsilon_b = \varepsilon_{a/4}$ and for a measurement with equal concentrations of the two species. Since the ACF for multiple non-interacting species is the sum of the individual ACFs weighted by square of their fractional intensity, the brighter species dominates the shape of the resulting ACF for the example shown here

amplitude decreases by a factor of 1.47 and the shape of the ACF, in this case, is dominated by the brighter species.

Signal coming from a fluorescent background can also be treated as an additional species. Typically, the background signal arises from a large number of weakly fluorescent molecules or from the detection of Rayleigh and Raman scattered light. In both of these cases, the ACF arising from the background "species" can be ignored. The former due to the large number of molecules $\gamma/(\mathrm{N})$ is small, the later because the photons arrive randomly, which does not lead to a correlation signal. However, to accurately determine the number of molecules from the fluorescence species of interest, the ACF has to be scaled by the square of the fraction intensity from the signal. This becomes important for quantitative measurements in the pM range, where the average number of signal molecules in the volume is less than one and the fraction intensity of the background becomes significant.

2.7 Cross-Correlation Spectroscopy

Fluorescence Correlation Spectroscopy can be expanded to investigate the interaction between molecules by using multiple detection channels. In addition to analyzing the ACF of the individual channels, a cross-correlation analysis between the different channels can be performed. A cross-correlation is detected when the signal fluctuates simultaneously in both channels, for example due to the fact that differently labeled proteins are diffusing as a complex. In Fluorescence Cross-Correlation Spectroscopy (FCCS), the cross-correlation function (CCF) is defined by:

$$G_{i\times j}(\tau) = \frac{\langle F_i(t)F_j(t+\tau)\rangle - \langle F_i(t)\rangle \langle F_j(t)\rangle}{\langle F_i(t)\rangle \langle F_j(t)\rangle}, \tag{31}$$

where the subscript $i\times j$ refers to the cross-correlation of the ith channel with the jth channel. In this article, we will not distinguish between $G_{i\times j}$ and $G_{j\times i}$ and in most cases, the CCF is symmetric and the two curves can be averaged together. Assuming identical, overlapping 3D Gaussian observation volumes, the CCF for a mixture of two species can be calculated (Schwille et al. 1997a, b, Kettling et al. 1998):

$$G_{1\times 2}(\tau) = \frac{\gamma \langle N_{12}\rangle}{\langle N_1 + N_{12}\rangle \langle N_2 + N_{12}\rangle} \left(\frac{1}{1+\tau/\tau_{D_{12}}}\right) \left(\frac{1}{1+(w_r/w_z)^2\tau/\tau_{D_{12}}}\right), \tag{32}$$

where $\langle N_{12}\rangle$ represents the average number of complexes detectable in both channels within the confocal observation volume. Whereas the amplitude of the ACF is inversely proportionally to the number of molecules, the CCF is proportional to the number of complexes containing both fluorophores. Hence, the CCF is a powerful means of determining whether two species interact. The usefulness of cross-correlation analysis was nicely shown by Petra Schwille and coworkers who used it to monitor the degradation of DNA by a restriction endonuclease EcoRI (Kettling et al. 1998).

2.8 Pulsed Interleaved Excitation

One difficulty in FCCS experiments is spectral crosstalk, where the fluorescence signal of one of the fluorophores is detected in the other channel. As FCCS is very sensitive to correlated signals and crosstalk is fully correlated, spectral crosstalk will be detected in the CCF. To eliminate spectral crosstalk, we developed pulsed interleaved excitation (PIE) (Müller et al. 2005). In PIE, two or more alternating or interleaved excitation sources are used and the information regarding which excitation source generated the detected photon is recorded. PIE can be used to remove spectral crosstalk in FCCS experiments (and in imaging experiments as well), it allows quantitative FCCS measurements with complexes that undergo FRET and provides information over the labeling stoichiometry in single molecule experiments. PIE is similar to the method of alternating laser excitation (ALEX), developed in the group of Shimon Weiss (Kapanidis et al. 2004, 2005, Lee et al. 2005), with the difference that we used subnanosecond lasers pulses and alternate the excitation source on the nanosecond time scale with repetition rates of 20–40 MHz. The high alternating frequency is necessary to allow submicrosecond resolution in FCS and FCCS measurements.

The principle of PIE is straightforward. In our realization of PIE, the excitation sources are synchronized to a master clock and the fluorescence signal is recorded with time-correlated single-photon-counting (TCSPC) detection. A schematic of the dual-color experimental setup is shown in Fig. 4a. For green and red excitation, picosecond pulsed diode lasers (at ~ 530 and 635 nm) were used. Diode lasers can be easily synchronized, making them ideal for use in PIE. From the arrival time

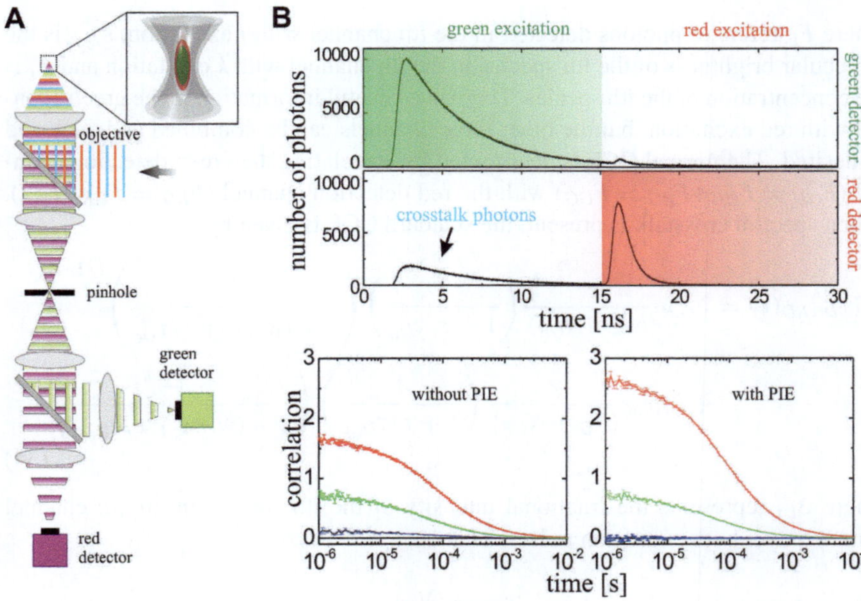

Fig. 4 The principle of PIE. **A** A schematic diagram of a dual-channel confocal microscope with pulsed interleaved excitation (PIE). The excitation pulses are shown in *blue* and *orange* and the resulting fluorescence is shown in *green* and *red* respectively. **B** *Upper panel*, The histogram of photon arrival times measured using time-correlated single-photon counting (TCSPC) with PIE for a sample of freely diffusing fluorophores, (Atto488 and Atto565). The green detection channel is plotted above and the red detection channel below. Photons arriving after green excitation occurring during the first 20 ns and photons arriving after red excitation are detected between 20 and 36 ns with respect to the master clock. By correlating photons detected in the green channel after green excitation (marked in *green*) with photons detected in the red channel after red excitation (marked in *red*), a crosstalk free FCCS function can be calculated. The resulting ACFs with and without PIE are shown in the *lower panel* for comparison

of photons with respect to the master clock, the appropriate excitation source is assigned to each detected photon. With the appropriate choice of excitation wavelengths and emission filters, spectral crosstalk is limited to the red detection channel with green excitation. The green channel with green excitation has only photons emitted from the green fluorophore and the photons detected in the red channel with red excitation are only emitted from the red fluorophore.

As PIE provides the additional information regarding which excitation source is responsible for each detected photon, photons can be categorized based on both the excitation source and detection channel. Hence, for a dual-color system, four channels are available:

$$\begin{aligned}
F_{GG}(t) &= \int dr\overline{W(\mathbf{r})}\left(\varepsilon_{G,G,G}C_G(\mathbf{r},t) + \varepsilon_{GR,G,G}C_{GR}(\mathbf{r},t)\right)\\
F_{GR}(t) &\approx 0\\
F_{RG}(t) &= \int dr\overline{W(\mathbf{r})}\left(\varepsilon_{G,R,G}C_G(\mathbf{r},t) + \varepsilon_{R,R,G}C_R(\mathbf{r},t) + \varepsilon_{GR,R,G}C_{GR}(\mathbf{r},t)\right)\\
F_{RR}(t) &= \int dr\overline{W(\mathbf{r})}\left(\varepsilon_{R,R,R}C_R(\mathbf{r},t) + \varepsilon_{GR,R,R}C_{GR}(\mathbf{r},t)\right)
\end{aligned} \tag{33}$$

where F_{ij} refers to photons detected in the ith channel with j excitation, $\varepsilon_{i,j,k}$ is the molecular brightness of the ith species in the jth channel with k excitation and C_i is the concentration of the ith species. There is no useful information in the green channel with red excitation, but the other three channels can be combined and analyzed as desired. The normal CCF is determined by correlating the green detection channel ($F_{GD} = F_{GG}+F_{RG} = F_{GG}$) with the red detection channel ($F_{RD} = F_{GR}+F_{RR}$). When spectral crosstalk is present, the standard CCF is given by[2]:

$$\begin{aligned}
G_{GD\times RD}(\tau) = {} & \Im_{G,R}\frac{\gamma}{\langle N_G + N_{GR}\rangle}\left(\frac{1}{1+\tau/\tau_{D_G}}\right)\left(\frac{1}{1+(w_r/w_z)^2\tau/\tau_{D_G}}\right)^{1/2}\\
& + \Im_{GR,R}\frac{\gamma}{\langle N_G + N_{GR}\rangle}\left(\frac{1}{1+\tau/\tau_{D_{GR}}}\right)\left(\frac{1}{1+(w_r/w_z)^2\tau/\tau_{D_{GR}}}\right)^{1/2}
\end{aligned} \tag{34}$$

where $\Im_{i,j}$ represents the fractional intensity of the ith species in the jth channel summed over both excitation pulses. That is:

$$\begin{aligned}
\Im_{G,R} &= \frac{\varepsilon_{G,G,R}\langle N_G\rangle}{\varepsilon_{G,G,R}\langle N_G + N_{GR}\rangle + \left(\varepsilon_{R,G,R} + \varepsilon_{R,R,R}\right)\langle N_R + N_{GR}\rangle} \quad \text{and}\\
\Im_{GR,R} &= \frac{\left(\varepsilon_{G,G,R} + \varepsilon_{G,R,R} + \varepsilon_{R,R,R}\right)\langle N_{GR}\rangle}{\varepsilon_{G,G,R}\langle N_G + N_{GR}\rangle + \left(\varepsilon_{R,G,R} + \varepsilon_{R,R,R}\right)\langle N_R + N_{GR}\rangle}.
\end{aligned} \tag{35}$$

[2]For simplicity, we assume that the quantum yields of the green and red fluorophores are identical in the single and double-labeled complexes. The more general equations can be found in Müller et al. *Biophys J* **89**:3508.

The detection of photons from the green fluorophore in the red channel after green excitation ($\varepsilon_{G,G,R}$) is, per definition, spectral crosstalk. Even when there are no interacting species present ($N_{GR} = 0$), the CCF has a finite amplitude due to spectral crosstalk as shown in Fig. 4b (blue curves) and can lead to misinterpretation of the data or decreased sensitivity.

As spectral crosstalk shows up in the red channel after green excitation, a spectral crosstalk-free cross correlation can be calculated by correlating photons detected in the green channel after green excitation with those detected in the red channel after red excitation:

$$G_{GG\times RR}(\tau) = \frac{\gamma \langle N_{GR}\rangle}{\langle N_G + N_{GR}\rangle \langle N_G + N_{GR}\rangle}\left(\frac{1}{1+\tau/\tau_{D_{GR}}}\right)\left(\frac{1}{1+(w_r/w_z)^2\tau/\tau_{D_{GR}}}\right)1/2. \tag{36}$$

The minimal detection level of interacting species is improved as shown in Fig. 4b. In addition, there is an increase in the amplitude of the red ACF when PIE is used. This happens because the crosstalk photons coming from green molecules, which appear as a second species in the ACF are removed. The amount of green photons in the red channel per a molecule is typically not large, but the effects can be significant under certain experimental conditions.

An additional advantage of PIE for FCCS experiments is the ability to perform quantitative measurements in the presence of FRET. Both FCCS and FRET are important techniques for investigating the interaction of different biomolecules. However, when FRET occurs between fluorophores in a FCCS measurement, a quantitative analysis is difficult. No sample is 100% labeled with photoactive donor and acceptor fluorophores. Hence, there are multiple species in the FCCS experiment. In the donor channel, there is the donor only species and the FRET species with a reduced molecular brightness. In the acceptor channel, there is the acceptor only species and the FRET species, which has an enhanced molecular brightness. The amplitude of the CCF in the presence of FRET is given by (Kohl et al. 2002, Müller et al. 2005):

$$G_{GD\times RD}(\tau) = \left(\frac{1-f_E}{1-f_E f_{GR,G}}\right)\left(\frac{1+f'_E}{1+f'_E f_{GR,R}}\right)\frac{\gamma N_{GR}}{\langle N_G + N_{GR}\rangle \langle N_R + N_{GR}\rangle}\cdot \left(\frac{1}{1+4D_{GR}\tau/\omega_r^2}\right)\left(\frac{1}{1+4D_{GR}\tau/\omega_s^2}\right)^{1/2} \tag{37}$$

Where

$$f_E = 1 - \varepsilon_{GR,G,G}/\varepsilon_{G,G,G}, \quad f'_E = \varepsilon_{GR,R,G}/\varepsilon_{R,R,R},$$
$$f_{GR,G} = \langle N_{GR}\rangle / (\langle N_G + N_{GR}\rangle)$$

and

$$f_{GR,R} = \langle N_{GR} \rangle / (\langle N_{GR} + N_R \rangle).$$

f_E is the FRET efficiency determined by the decrease in molecular brightness of the donor and f'_E is the increase in sensitized emission of the acceptor due to FRET and is proportional to the FRET efficiency. The proportionality constant depends upon the intensity of the laser used for direct acceptor excitation. A quantitative analysis of the number of species diffusing with both labels, N_{GR}, requires knowledge of the above parameters.

PIE can be used to perform quantitative FCCS measurements when the direct excitation of the acceptor with green light is negligible. In this case, the photons detected in the red channel after green excitation are due to FRET or from spectral crosstalk of the donor. To perform a quantitative FCCS analysis in the presence of FRET, the photons that are lost due to FRET and crosstalk are added back to the donor channel. In other words, we correlate the photons detected after *green excitation* ($F_{GX} = F_{GG} + F_{GR}$) with the photons detected after *red excitation* ($F_{RX} = F_{RG} + F_{RR} \approx F_{RR}$). The CCF reverts to Eq. (32), even in the presence of FRET.

In addition to being able to perform quantitative FCCS experiments in the presence of FRET, it is possible to determine the average FRET efficiency of the sample using FCCS. Widengren and coworkers were the first to develop methods for determining the static FRET efficiency from FCS experiments (Widengren et al. 2001). With the aid of PIE, the FRET efficiency can be determined from the ratio of auto- and cross-correlation functions as given in Eq. (38),

$$f_E = \frac{1 - \dfrac{G_{GR \times RR}(0)}{G_{GX \times RR}(0)}}{1 - \dfrac{G_{GG \times RR}(0)}{G_{RR \times RR}(0)}}. \tag{38}$$

The elegance of this approach is that the labeling efficiency of the donor molecules as well as the FRET efficiency are determined directly from the FCCS measurement. No calibration is necessary, provided the assumptions regarding the different probe volumes and detection correction factor are correct. Hence, with PIE it is possible to perform artifact-free quantitative cross correlation analyses in the presence of spectral crosstalk and FRET.

PIE can be used in conjunction with many different fluorescence methods such as confocal imaging, FCCS, and spFRET to improve them or enhance their capabilities. For example, PIE can be used together with burst analysis to investigate the heterogeneity of the sample and visualize subpopulations. As was first shown with ALEX (Kapanidis et al. 2004), PIE can be used to determine the stoichiometry, S, of the sample, defined as:

$$S = \frac{F_{CG} + F_{RG}}{F_{GG} + F_{RG} + F_{RR}}. \tag{39}$$

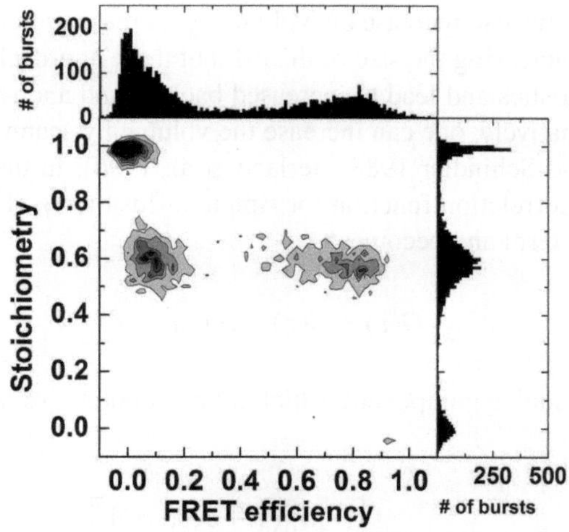

Fig. 5 FRET and stoichiometry. A two-dimensional plot of stoichiometry versus FRET efficiency is shown for a mixture of 40 bp DNAs labelled with Atto532 and Atto647N at separations of 9 and 23 base pairs. The stoichiometry information allows DNAs with only a single active fluorophore, either donor or acceptor, to be distinguished from double-labelled DNA molecules. One-dimensional histograms of the FRET efficiency and stoichiometry are shown *above* and to the *right* of the plot for comparison

The stoichiometry factor opens a new dimension, making it possible to determine the labeling ratio of donor to acceptor. Measurements from a mixture of DNA molecules labeled at two different distances with a FRET pair are shown in Fig. 5. The peak observed with a high stoichiometry value and low FRET efficiency comes from the donor only species. Molecules that have a low stoichiometry value are acceptor-only complexes. Without a donor molecule, the FRET efficiency is undefined, but the calculated FRET efficiency is typically high as most photons come from direct excitation of the acceptor molecule. Molecules with an intermediate stoichiometry correspond to molecules (DNA in this case) that have both an active donor and acceptor. The ability to separate the donor-only and acceptor-only complexes makes it possible to distinguish and accurately determine low FRET efficiency conformations. More detailed information over PIE, calculations for non-identical probe volumes, signal-to-noise considerations, and different sensitivities to donor and acceptor molecules are given in (Müller et al. 2005).

3 Advanced FCS Methods

3.1 Scanning FCS

In many cases, one has to deal with strongly diluted samples and/or slowly diffusing particles. These conditions require very long acquisition times in order to obtain reasonable statistics over the fluctuations. To improve the probability of detecting a particle, one can increase the volume by increasing the size of the confocal pinhole and/or increasing the size of the psf, but these approaches give marginal improvement in statistics and lead to increased background and reduced molecular brightnesses. Alternatively, one can increase the volume by scanning the sample or the laser (Meyer and Schindler 1988, Berland et al. 1996). In the case of circular scanning FCS, the correlation function for simple diffusion G_D (Eq. 6) acquires an additional scanning term and becomes:

$$G(\tau) = S(\tau) \cdot G_D(\tau) \tag{40}$$

where $S(\tau)$ is the scanning component, which for one photon excitation is expressed as:

$$S(\tau) = \exp\left[-\frac{4r_0^2\left(1-\cos(\omega\tau)\right)}{\left(w_r^2+4D\tau\right)}\right] \tag{41}$$

with r_0 being the radius of the orbit and $\omega = 2\pi\nu$ is the circular frequency of the scan (Berland et al. 1996). A comparison of scanning FCS and standard FCS is shown in Fig. 6a.

Scanning FCS is advantageous for experiments in living cells where diffusion is often slow and photobleaching is a problem. By scanning, one increases the effective volume, allowing multiple measurements to be performed quasi simultaneously and the results can be averaged together to provide better statistics. In addition, the excitation light is distributed within the cell, leading to a decrease in the photobleaching rate. A scanning FCS curve of Alexa 488-labeled dsDNA diffusing in aqueous solution is shown in Fig. 6a. A scanning rate of 1 kHz and scanning radii of 250 nm and 1 μm were used. For comparison, experiments without scanning were also performed. The resulting ACFs show the modulation described by the Eq. (42). At early times, the ACF decreases rapidly because the observation volume is being scanned to a new position where there is no correlation, as the molecules diffuse much slower than the rate of scanning. The correlation increases again as the beam returns to the original position of the orbit where the particle was initially detected. Hence, spikes are detected at an integral number of periods and the amplitude of the oscillations corresponds to the amplitude of the ACF measured without scanning. When the radius is large enough that the observation volumes from different regions of the scan do not overlap, the correlation drops to zero, at least at early times where the particle cannot diffuse significantly. This occurs because the observation volume is scanned much more rapidly than the molecules can diffuse and non-overlapping regions along the scan contain different molecules and, therefore, do not correlate. After a complete period, the original molecules will again be detected and hence, a correlation will be observed. As the correlation time increases and molecules have time to explore a larger region of the sample, the spikes broadened, the amplitude decreases and the minimum will no longer go to zero. For the scan radius of 250 nm, the ACF does not drop to zero as the scan is comparable to the size of the psf and

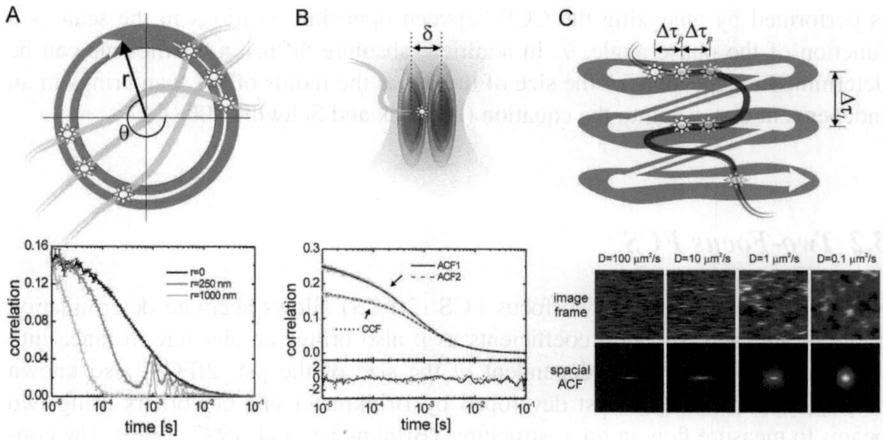

Fig. 6 Advanced FCS techniques. **a** Scanning FCS. *Upper panel*, a schematic of a scanning FCS experiment where the observation volume is rotated in a circle. *Lower panel*, ACFs of DNA labelled with Atto488 diffusing in aqueous solution are shown for two different scanning radii at a frequency of 1 kHz, one with a radius of 250 nm and one with a radius of 1 μm. An ACF without scanning is shown for comparison. When the radius is larger than the psf, the ACF goes to zero between the harmonics of the scanning frequency. **b** *Upper panel*. A schematic of a two-focus FCS experiment is shown where two observations volumes are measured and correlated. *Bottom panel*. The ACFs from each observation volume and CCF of the two volumes along with a global fit to the data are shown for fluorescently labelled myoglobin freely diffusing in buffer at pH 3.2. The distance between the two volumes, δ, brings an independent dimension into the analysis, making it possible to accurately determine the diffusion coefficient of the labelled molecule as well as the size of the observation volume. **c** *Upper panel*. Schematic of a raster scan. A raster scan has typically two axes collected with different scan rates, one dimension, e.g. *x*, is scanned quickly with a dwell time of τ_p per pixel and the second axis, e.g. *y*, is scanned more slowly with a line scan speed of τ_l. *Lower panel*. From the raster scan images, a two-dimensional correlation function is calculated. Simulated data is shown for four different diffusion coefficients. Fast diffusing objects will have a correlation function elongate along the *x* dimension while slower moving objects will be more symmetric (this panel is adapted from Brown et al. 2008)

hence, there is still a significant overlap of the observation volumes at all positions during the scan.

Scanning FCS can also be done using line scans. As most commercially available scanning microscopes are not capable of scanning in a circle, line-scanning FCS provides a good alternative. The power of line-scanning FCS has been demonstrated by the group of Petra Schwille where they investigated the diffusion of molecules in membranes (Ries and Schwille 2006, Ries et al. 2009). The combination of slowly diffusing molecules and fluctuations of the membrane make FCS experiments in membranes challenging. By using the image information available in the line scan, the authors could correct for motion of the membrane and then perform the FCS analysis.

In scanning FCS, spatial information is also available in the raw data. Hence, it is possible to detect both the direction and rate of flow when it is present. This is performed by analyzing the CCF between opposing positions in the scan as a function of the phase angle, θ. In addition, absolute diffusion coefficients can be determined independent of the size of the psf as the radius of the scan brings in an independent distance into the equation (Petrasek and Schwille 2008).

3.2 Two-Focus FCS

Similar to scanning FCS, two-focus FCS (2fFCS) allows accurate determination of flow rates and diffusion coefficients as it also brings an absolute distance into the measurement that is independent of the size of the psf. 2fFCS, also known as dual-beam FCS, was first developed by Brinkmeier and coworkers using two beams to measure flow in microstructures (Brinkmeier et al. 1997, 1999). The concept of 2fFCS is shown schematically in Fig. 6b. The excitation laser was split into two beams and focused through the same objective to form two volumes separated by a couple of micrometers. The emitted fluorescence is collected through the same objective and focused on separate detectors. Assuming two identical three-dimensional Gaussian excitation profiles, the cross-correlation function can be calculated analytically and is given by:

$$G(\tau,\delta)=\frac{\gamma}{N}\left[\frac{1}{1+\tau/\tau_D}\right]\cdot\left[\frac{1}{1+\left(\omega_r/\omega_z\right)^2\tau/\tau_D}\right]^{1/2}\exp\left[-\frac{\delta^2}{\omega_r^2\left(1+\tau/\tau_D\right)}\right], \tag{42}$$

where δ is the distance between the centers of the two volumes. Molecules diffusing through the centers of both confocal volumes will cause a maximal cross-correlation amplitude whereas molecules that cross perpendicular to the axis that separates the two volumes will result in minimal cross-correlation.

An elegant implementation of 2fFCS was performed by Dertinger and colleagues (2007). This method uses a pulsed laser with a repetition rate of several MHz (typically 40 MHz) and a Nomarski prism. The output of the laser is split into two equal but orthogonal polarizations. One polarization is fed into an optical delay line (e.g. a fiber optics cable) before being recombined into the same excitation pathway. Before the objective, the Nomarski prism laterally displaces the two polarizations into separate beams, forming the two foci, which are separated by 150–500 nm. The spacing depends on the particular combination of the objective and the prism and the wavelength of the excitation light. The fluorescence is then collected via the same objective, passed through a pinhole to allow confocal detection and focused on a single detector. The confocal pinhole is typically chosen large enough to allow detection of in-focus light from both excitation volumes. As the volumes are spatially overlapping, the principle of PIE

(Müller et al. 2005) was used to determine from which focus the fluorescence was generated.[3]

The Gaussian approximation for a diffraction-limited psf is acceptable in the x–y plane, but is often inadequate in the axial direction. In the work by Dertinger et al. (2007), a two-dimensional Gaussian was used to approximate the psf in x and y and an empirical function was used to describe the axial dependence of the psf. In addition, the width of the two-dimensional Gaussian was dependent on the z-position. The psf was given by:

$$W(\mathbf{r}) = \frac{\kappa(z)}{w(z)} \exp\left[-\frac{2\left(x^2+y^2\right)}{w(z)}\right] \tag{43}$$

where

$$\begin{aligned} w(z) &= w_0\sqrt{1+\left(\frac{\lambda_{ex}z}{\pi w_0^2 n}\right)^2} \\ \kappa(z) &= 1-\exp\left(-\frac{2a^2}{R^2(z)}\right) \\ R(z) &= R_0\sqrt{1+\left(\lambda_{em}z/\pi R_0^2 n\right)^2} \end{aligned}, \tag{44}$$

w_0 and n are the beam waist and the refractive index of immersion medium (typically water), λ_{ex} is the excitation wavelength, λ_{em} is the emission wavelength, a is the radius of the confocal pinhole divided by the magnification, and R_0 is a model parameter.

For the above psf, the cross-correlation function has to be solved numerical and is given by (Dertinger et al. 2007).

$$\begin{aligned} G(\tau,\delta) = \frac{1}{c}\sqrt{\frac{16}{\pi^3 D\tau}}\Bigg[\int_{-\infty}^{\infty} dz_1 \int_{-\infty}^{\infty} dz_2 \frac{\kappa(z_1)\kappa(z_2)}{8Dt+w^2(z_1)+w^2(z_2)} \\ \times \exp\left[-\frac{(z_2-z_1)^2}{4Dt}-\frac{2\delta^2}{8Dt+w^2(z_1)+w^2(z_2)}\right]\Bigg]\Bigg/\left[\int_{-\infty}^{\infty} k(z)dz\right]^2, \end{aligned} \tag{45}$$

where c is the concentration of molecules and D is the diffusion coefficient.

[3]Polarized detection alone would not be sufficient to separate the two volumes as the detected fluorescence is strongly depolarized due to rotation of the fluorophores on the time scale of the fluorescence lifetime.

3.3 Raster Image Correlation Spectroscopy

Another adaptation of FCS that is particularly advantageous for experiments conducted in live cells is Raster Image Correlation Spectroscopy (RICS) (Digman et al. 2005a, b). As a scanning method, RICS gathers statistics over a larger volume providing good statistics with minimal photobleaching and, in addition, provides image information (Brown et al. 2008, Digman et al. 2008). From the RICS data, it is possible to use the image information to correct for slow cellular movements. RICS uses the spatiotemporal information inherent in a raster scan to extract information regarding the average number and mobility of fluorescent molecules. As the name implies, it analyzes raster-scanned images and can be performed with the latest-generation commercial confocal microscopes (e.g. Zeiss LSM 780 or Nikon A1). The scanned images must be of sufficient linearity to allow averaging of the correlation over the selected scan region and the characteristics of the scanner (pixel size, dwell time per pixel and time between consecutive line scans) are known. The resulting raster scan images can be analyzed *post facto* using home written or commercially available software (for example SimFCS, www.lfd.uci.edu).

In a raster-scanned image, the focus of the microscope is scanned across the sample in a saw-tooth pattern with rapid line scans along one axis (e.g. x) and a slower frame scan along the second axis (e.g. y) (Fig. 6c). Due to the well-defined scanning pattern, there is a correlation between the position of a pixel and time it was measured (provided the scanning speed is constant and spacing is linear). This correlation is visible in the two-dimensional ACF of the raster-scanned image. In RISC, the ACF is given by the contributions from scanning (S) and from diffusion of the molecules (G_D):

$$G_s(\xi, \psi) = S(\xi, \psi) \cdot G_D(\xi, \psi), \tag{46}$$

where

$$S(\xi, \psi) = \exp\left[- \frac{\left(\frac{|\xi|\delta x}{w_r}\right)^2 + \left(\frac{|\psi|\delta y}{w_r}\right)^2}{1 + \frac{4D(\tau_p|\xi| + \tau_l|\psi|)}{w_r^2}} \right], \tag{47}$$

$$G_D(\xi, \psi) = \frac{\gamma}{N}\left(1 + \frac{4D\left(\tau_p\,|\xi| + \tau_l\,|\psi|\right)}{w_r^2}\right)^{-1}\left(1 + \frac{4D\left(\tau_p\,|\xi| + \tau_l\,|\psi|\right)}{w_z^2}\right)^{-1/2}, \tag{48}$$

ξ and ψ are the spatial lags in x and y, δx and δy are the pixel dimensions, τ_p and τ_l are the pixel dwell time and interline delay, respectively. From the practical point of view, it is necessary to oversample the psf by a factor of ~10, which results in pixel dimensions of around 20–50 nm, pixel dwell time τ_p can vary depending on the particular application from 2 to 100 μs. Image size varies typically from 32 × 32 to 1,024 × 1,024 pixels.

As the scan speed along the two axes is different, RICS is sensitive to a broad range of diffusion coefficients. For a quickly diffusing molecule, it will only be detected in a few adjacent pixels within a single line scan. Upon returning to the same x coordinates in the adjacent line, the particle will have diffused elsewhere. Hence, the RICS ACF will be very narrow (Fig. 6c). On the other hand, slowly moving particles will have a high probability of being detected in the adjacent line scans and the ACF will become broader. For immobilized particles that are significantly smaller than the size of the psf of the microscope, the RICS ACF will simply be the square of the psf. Hence, a RICS analysis of small beads (e.g. 20 nm ø) immobilized on the surface can be a excellent tool for determining the size of the psf of a particular setup. As RICS is based upon raster-scanned images, the image information is available in the data and can be used, for example, to correct for artifacts such as cell migration or drift.

When performing raster imaging with two detection channels, a cross-correlation analysis can also be performed (as for normal FCS) to investigate the interaction between molecules labeled with different color fluorophores (Digman et al. 2009a, b).

References

Aragón, S.F. and Pecora, R. (1976) Fluorescence correlation spectroscopy as a probe of molecular dynamics. *J. Chem. Phys.* **64**: 1791–8103.

Berland, K.M., So, P.T., Chen, Y., Mantulin, W.W. and Gratton, E. (1996) Scanning two-photon fluctuation correlation spectroscopy: particle counting measurements for detection of molecular aggregation. *Biophys. J.* **71**: 410–420.

Bismuto, E., Gratton, E. and Lamb, D.C. (2001) Dynamics of ANS binding to tuna apomyoglobin measured with fluorescence correlation spectroscopy. *Biophys. J.* **81**: 3510–3521.

Bonnet, G., Krichevsky, O. and Libchaber, A. (1998) Kinetics of conformational fluctuations in DNA hairpin-loops. *Proc. Natl. Acad. Sci. USA* **95**: 8602–8606.

Boukobza, E., Sonnenfeld, A. and Haran, G. (2001) Immobilization in surface-tethered lipid vesicles as a new tool for single biomolecule spectroscopy. *J. Phys. Chem. B* **105**: 12165–12170.

Brinkmeier, M., Dorre, K., Riebeseel, K. and Rigler, R. (1997) Confocal spectroscopy in microstructures. *Biophys. Chem.* **66**: 229–239.

Brinkmeier, M., Dorre, K., Stephan, J. and Eigen, M. (1999) Two-beam cross-correlation: a method to characterize transport phenomena in micrometer-sized structures. *Anal. Chem.* **71**: 609–616.

Brown, C.M., Dalal, R.B., Hebert, B., Digman, M.A., Horwitz, A.R. and Gratton, E. (2008) Raster image correlation spectroscopy (RICS) for measuring fast protein dynamics and concentrations with a commercial laser scanning confocal microscope. *J. Microsc.* **229**: 78–91.

Brown, R.H. and Twiss, R.Q. (1956) Correlation between photons in two coherent beams of light. *Nature* **177**: 27–29.

Dertinger, T., Pacheco, V., von der Hocht, I., Hartmann, R., Gregor, I. and Enderlein, J. (2007) Two-focus fluorescence correlation spectroscopy: a new tool for accurate and absolute diffusion measurements. *Chemphyschem* **8**: 433–443.

Dickson, R.M., Cubitt, A.B., Tsien, R.Y. and Moerner, W.E. (1997) On/off blinking and switching behaviour of single molecules of green fluorescent protein. *Nature* **388**: 355–358.

Digman, M.A., Brown, C.M., Sengupta, P., Wiseman, P.W., Horwitz, A.R. and Gratton, E. (2005a). Measuring fast dynamics in solutions and cells with a laser scanning microscope. *Biophys. J.* **89**: 1317–1327.

Digman, M.A., Sengupta, P., Wiseman, P.W., Brown, C.M., Horwitz, A.R. and Gratton, E. (2005b). Fluctuation correlation spectroscopy with a laser-scanning microscope: exploiting the hidden time structure. *Biophys. J.* **88**: L33–36.

Digman, M.A., Dalal, R., Horwitz, A.F. and Gratton, E. (2008) Mapping the number of molecules and brightness in the laser scanning microscope. *Biophys. J.* **94**: 2320–2332.

Digman, M.A., Wiseman, P.W., Choi, C., Horwitz, A.R. and Gratton, E. (2009a). Stoichiometry of molecular complexes at adhesions in living cells. *Proc. Natl. Acad. Sci. USA* ***106***: 2170–2175.

Digman, M.A., Wiseman, P.W., Horwitz, A.R. and Gratton, E. (2009b). Detecting protein complexes in living cells from laser scanning confocal image sequences by the cross correlation raster image spectroscopy method. *Biophys. J.* **96**: 707–716.

Ehrenberg, M. and Rigler, R. (1974) Rotational brownian motion and fluorescence intensity fluctuations. *Chem. Phys.* **4**: 390–401.

Ehrenberg, M. and Rigler, R. (1976) Fluorescence correlation spectroscopy applied to rotational diffusion of macromolecules. *Q. Rev. Biophys.* **9**: 69–81.

Eigen, M. and Rigler, R. (1994) Sorting single molecules: application to diagnostics and evolutionary biotechnology. *Proc. Natl. Acad. Sci. USA* **91**: 5740–5747.

Einstein, A. (1905) Über die von der molekularkinetischen Theorie der Wärme geforderte Bewegung von in ruhenden Flüssigkeiten suspendierten Teilchen. *Ann. Phys.* **17**: 549–560.

Elson, E.L. and Magde, D. (1974) Fluorescence correlation spectroscopy. I. Conceptual basis and theory. *Biopolymers* **13**: 1–27.

Felekyan, S., Kuhnemuth, R., Kudryavtsev, V., Sandhagen, C., Becker, W. and Seidel, C.A.M. (2005) Full correlation from picoseconds to seconds by time-resolved and time-correlated single photon detection. *Rev. Sci. Instrum.* **76**: 083104–083114.

Ha, T., Rasnik, I., Cheng, W., Babcock, H.P., Gauss, G.H., Lohman, T.M. and Chu, S. (2002) Initiation and re-initiation of DNA unwinding by the *Escherichia coli* Rep helicase. *Nature* **419**: 638–641.

Heimstadt, O. (1911) Das Fluoreszenzmicroskop. *Z. wiss. Mikrosk.* **28**: 330.

Kapanidis, A.N., Lee, N.K., Laurence, T.A., Doose, S., Margeat, E. and Weiss, S. (2004). Fluorescence-aided molecule sorting: analysis of structure and interactions by alternating-laser excitation of single molecules. *Proc. Natl. Acad. Sci. USA* **101**: 8936–8941.

Kapanidis, A.N., Laurence, T.A., Lee, N.K., Margeat, E., Kong, X. and Weiss, S. (2005) Alternating-laser excitation of single molecules. *Acc. Chem. Res.* ***38***: 523–533.

Kask, P., Piksarv, P., Pooga, M., Mets, Ü. and Lippmaa, E. (1989) Separation of the rotational contribution in fluorescence correlation experiments. *Biophys. J.* **55:** 213–220.

Kettling, U., Koltermann, A., Schwille, P. and Eigen, M. (1998) Real-time enzyme kinetics monitored by dual-color fluorescence cross-correlation spectroscopy. *Proc. Natl. Acad. Sci. USA* **95:** 1416–1420.

Kohl, T., Heinze, K.G., Kuhlemann, R., Koltermann, A. and Schwille, P. (2002) A protease assay for two-photon crosscorrelation and FRET analysis based solely on fluorescent proteins. *Proc. Natl. Acad. Sci. USA* **99:** 12161–12166.

Koppel, D.E. (1974) Statistical accuracy in fluorescence correlation spectroscopy. *Phys. Rev. A.* **10**: 1938–1945.

Koppel, D.E., Axelrod, D., Schlessinger, J., Elson, E.L. and Webb, W.W. (1976) Dynamics of fluorescence marker concentration as a probe of mobility. *Biophys. J.* **16**: 1315–1329.

Lamb, D.C., Schenk, A., Röcker, C. and Nienhaus, G.U. (2000a). Determining chemical rate coefficients using time-gated fluorescence correlation spectroscopy. *J. Phys. Org. Chem.* **13:** 654–658.

Lamb, D.C., Schenk, A., Röcker, C., Scalfi-Happ, C. and Nienhaus, G.U. (2000b). Sensitivity enhancement in fluorescence correlation spectroscopy of multiple species using time-gated detection. *Biophys. J.* **79**: 1129–1138.

Lee, N.K., Kapanidis, A.N., Wang, Y., Michalet, X., Mukhopadhyay, J., Ebright, R.H. and Weiss, S. (2005) Accurate FRET measurements within single diffusing biomolecules using alternating-laser excitation. *Biophys. J.* **88**: 2939–2953.

Lehman, H. (1913) Das Lumineszenzmicroscop. *Zeitschrift für Wissenschaftliche Microskopie* **30**: 417–470.

Lu, H.P., Xun, L. and Xie, X.S. (1998) Single-molecule enzymatic dynamics. *Science* **282**: 1877–1882.

Magde, D., Elson, E.L. and Webb, W.W. (1972) Thermodynamic fluctuations in a reacting system – measurement by fluorescence correlation spectroscopy. *Phys. Rev. Lett.* **29**: 705–708.

Magde, D., Elson, E.L. and Webb, W.W. (1974) Fluorescence correlation spectroscopy. II. An experimental realization. *Biopolymers* **13**: 29–61.

Magde, D. (1976) Chemical kinetics and fluorescence correlation spectroscopy. *Q. Rev. Biophys.* **9**: 35–47.

Magde, D., Webb, W.W. and Elson, E.L. (1978) Fluorescence correlation spectroscopy. III. Uniform translation and laminar flow. *Biopolymers* **17**: 361–376.

Meyer, T. and Schindler, H. (1988) Particle counting by fluorescence correlation spectroscopy. Simultaneous measurement of aggregation and diffusion of molecules in solutions and in membranes. *Biophys. J.* **54**: 983–993.

Müller, B.K., Zaychikov, E., Bräuchle, C. and Lamb, D.C. (2005) Pulsed interleaved excitation. *Biophys. J.* **89**: 3508–3522.

Müller, C.B., Loman, A., Pacheco, V., Koberling, F., Willbold, D., Richtering, W. and Enderlein, J. (2008) Precise measurement of diffusion by multi-color dual-focus fluorescence correlation spectroscopy. *EPL* **83**: 46001.

Petrasek, Z. and Schwille, P. (2008) Precise measurement of diffusion coefficients using scanning fluorescence correlation spectroscopy. *Biophys. J.* **94**: 1437–1448.

Rauer, B., Neumann, E., Widengren, J. and Rigler, R. (1996) Fluorescence correlation spectrometry of the interaction kinetics of tetramethylrhodamin a-bungarotoxin with Torpedo californica acetylcholine receptor. *Biophys. Chem.* **58**: 3–12.

Reichert, K. (1911) Das Fluorescenczmikroskop. *Phys. Z.* **12**: 1010–1011.

Rhoades, E., Gussakovsky, E. and Haran, G. (2003) Watching proteins fold one molecule at a time. *Proc. Natl. Acad. Sci. USA* **100**: 3197–3202.

Ries, J. and Schwille, P. (2006) Studying slow membrane dynamics with continuous wave scanning fluorescence correlation spectroscopy. *Biophys. J.* **91**: 1915–1924.

Ries, J., Chiantia, S. and Schwille, P. (2009) Accurate determination of membrane dynamics with line-scan FCS. *Biophys. J.* **96**: 1999–2008.

Rigler, R., Kask, P., Mets, Ü. and Widengren, J. (1993) Fluorescence correlation spectroscopy with high count rate and low background: analysis of translational diffusion. *Eur. Biophys. J.* **22**: 169–175.

Schwille, P., Bieschke, J. and Oehlenschlager, F. (1997a). Kinetic investigations by fluorescence correlation spectroscopy: the analytical and diagnostic potential of diffusion studies. *Biophys. Chem.* **66**: 211–228.

Schwille, P., MeyerAlmes, F.J. and Rigler, R. (1997b). Dual-color fluorescence cross-correlation spectroscopy for multicomponent diffusional analysis in solution. *Biophys. J.* **72**: 1878–1886.

Svedberg, T. and Inouye, K. (1911) Eine neue Methode zur Prüfung der Gültigkeit des Boyle-Gay-Lussacschen Gesetzes für Kolloide Lösungen. *Z. Phys. Chem.* **77**: 145–191.

Thompson, N.L., Burghardt, T.P. and Axelrod, D. (1981) Measuring surface dynamics of biomolecules by total internal reflection fluorescence with photobleaching recovery or correlation spectroscopy. *Biophys. J.* **33**: 435–454.

Thompson, N.L. (1991) Fluorescence correlation spectroscopy. In: *Topics in fluorescence spectroscopy, volume 1: techniques*, J.R. Lakowicz, ed. Plenum Press, New York, NY, pp. 337–378.

Torres, T. and Levitus, M. (2007) Measuring conformational dynamics: a new FCS-FRET approach. *J. Phys. Chem. B* **111**: 7392–7400.

von Smoluchowski, M. (1906) Zur kinetischen Theorie dier Brownschen Molekularbewegung und der Suspensionen. *Ann. Phys.* **21**: 756–780.

Webb, R.H. (1996) Confocal optical microscopy. *Rep. Progr. Phys.* **59**: 427.

Wennmalm, S., Edman, L. and Rigler, R. (1997) Conformational fluctuations in single DNA molecules. *Proc. Natl. Acad. Sci. USA* **94**: 10641–10646.
Widengren, J., Rigler, R. and Mets, Ü. (1994) Triplet-state monitoring by fluorescence correlation spectroscopy. *J. Fluoresc.* **4**: 255–258.
Widengren, J., Mets, Ü. and Rigler, R. (1995) Fluorescence correlation spectroscopy of triplet states in solution: a theoretical and experimental study. *J. Phys. Chem.* **99**: 13368–13379.
Widengren, J., Schweinberger, E., Berger, S. and Seidel, C.A.M. (2001) Two new concepts to measure fluorescence resonance energy transfer via fluorescence correlation spectroscopy: theory and experimental realizations. *J. Phys. Chem. A* **105**: 6851–6866.

Time-Resolved FT-IR Spectroscopy for the Elucidation of Protein Function

Michael Schleeger, Ionela Radu, and Joachim Heberle

Abstract Time-resolved Fourier transform infrared spectroscopy (FT-IR) has been proven to be an excellent method with important applications in bioscience. In particular, it is possible to monitor the temporal evolution of the reaction mechanism of complex machineries as membrane proteins, where other techniques encounter significant experimental difficulties. Here, we summarize the classical principles and experimental realizations of time-resolved FT-IR spectroscopy together with new developments realized in our laboratory. Examples from applications to retinal proteins are reviewed that showcase the impact of time-resolved FT-IR spectroscopy on the understanding of protein reactions on the level of single bonds.

Keywords Step-scan spectroscopy · Vibrational spectroscopy · Retinal proteins · Microfluidics · Heme proteins

1 Introduction

One crucial task of spectroscopy in bioscience is to identify the structure of proteins on an atomic level. To perform their biological function, proteins may change between several related conformations. Such changes can be induced by binding of substrates to indicative sites of a protein and/or specific interactions with cofactors and bound ligands.

All intra- and intermolecular alterations sensitively modulate the infrared spectrum of proteins because the vibrational modes are basically determined by the structure. During its activity only small parts of the protein undergo molecular changes. Their identification from the infrared spectrum is impeeded by the strong background absorbance of the whole protein. This difficulty is elegantly resolved by

J. Heberle (✉)
Experimental Molecular Biophysics, Department of Physics, Free University of Berlin, D-14195 Berlin, Germany
e-mail: jheberle@zedat.fu-berlin.de

J. Brnjas-Kraljević, G. Pifat-Mrzljak (eds.), *Supramolecular Structure and Function 10*, DOI 10.1007/978-94-007-0893-8_2,

forming the IR difference between two stable reaction states, an ingenious method that makes IR spectroscopy selective. The difference spectrum exhibits bands characteristic only for those molecular groups which undergo a change during protein activity. As the resulting absorbance changes are usually only in the order of 10^{-4} or less, their proper detection requires sensitive instrumentation. This requirement is fullfilled by Fourier-Transform (FT) techniques which are currently applied in studies of protein structure and functionality.

Insights into the mechanism of protein function comprises to resolve the evolution of intermediate states along the reaction pathway in a non-invasive manner. For this purpose, time-resolved FT-IR spectroscopy has proven to be superior to many other spectroscopies. Single-wavelength techniques using pulsed pump-probe or continuous wave (cw) lasers are applied to interrogate dynamic processes with ultrafast time scales up to femtoseconds. Although these methods offer high sensitivity, they do not take advantage of the multiplex advantage of FT-IR spectroscopy. During the past three decades, two FT-IR techniques showed growing impact on time-dependent investigations: the rapid-scan and the step-scan technique (Manning et al. 1991, Uhmann et al. 1991). The latter achieves a time resolution from microseconds to nanoseconds.

In the present article we will provide an introduction to the principles of time-resolved FT-IR spectroscopy along with some instrumental considerations. After illustrating how time-resolved FT-IR spectroscopy was applied to the elucidation of the mechanism of the proton pump bacteriorhodopsin, we will finally present new technical advances to overcome some restictions of the conventional methodologies.

2 Time-Resolved FT-IR Techniques

2.1 Rapid-Scan FT-IR Spectroscopy

The central element of an FT-IR spectrometer is the interferometer. A typical FT-IR setup is schematically shown in Fig. 1. The infrared light emitted from a globar IR source passes an aperture which controls the amount of IR light probing the sample. The beam enters the interferometer where the spectral encoding takes place. Next, the light passes through the sample and is focused on a semiconductor detector (MCT, mercury cadmium telluride). The detected signal represents an interference pattern and is known as the interferogram. The output signal of the detector is amplified, digitised and transfered into a personal computer. There the infrared spectrum is computed by means of Fourier transform (spectral decoding).

The central challenge of kinetic FT-IR spectroscopy is the achievement of high time resolution. An obvious approach is to record the spectra uninterruptedly while the observed temporal process is accomplished. In rapid-scan FT-IR the moving mirror of the interferometer rapidly scans back and forth. The experiment starts by collecting a reference interferogram $I_{t0}(x)$, which is necessary to form difference spectra. After a certain delay the reaction is initiated for instance by a laser pulse

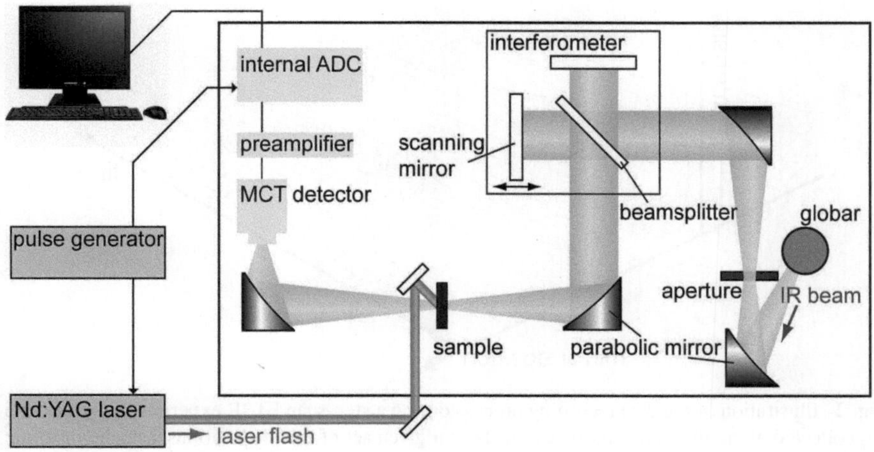

Fig. 1 Experimental setup used in time-resolved step-scan FT-IR spectroscopy

and the time-resolved interferograms $I_{ti}(x)$ are recorded one after another. In order to improve the signal-to-noise ratio the experiment can be repeated many times. After completing the measurement the interferograms are Fourier-transformed (Griffiths and de Haseth 1986) to yield the reference $S_{t0}(\nu)$ and the time-resolved spectra $S_{ti}(\nu)$, respectively. Finally, the absorption changes during the reaction cycle are calculated by the relation $\Delta A_{ti}(\nu) = -\log(S_{ti}(\nu)/S_{t0}(\nu))$.

It is evident that the time resolution is determined by the velocity of the moving mirror in rapid-scan experiments. Moreover, the time-resolution depends on the optical resolution because the better the spectral resolution, i.e. the longer the scan length, the worse the time resolution will be at a given velocity of the movable mirror. As an example of typical parameters used in the spectroscopy of protein samples, the time-dependent phenomena can be resolved to 10 ms with a scanning velocity of 10.1 cm·s^{-1} and a spectral resolution of 4 cm^{-1}.

2.2 *Step-Scan FT-IR Spectroscopy*

Step-scan FT-IR spectroscopy exhibits no mechanical restrictions while acquiring time-resolved data and maintains the benefits of the multiplex advantage of the continuous-scan FT-IR spectroscopy. During a step-scan experiment, the movable mirror moves step-wise from one sampling position to the next (Fig. 2 vertical lines). At each position the mirror is held fixed and the kinetic process is initiated, commonly by a laser pulse. A transient recorder digitizes the intensity changes over time. Once the process is completed, the mirror is stepped to the next position and the reaction is initiated again. The kinetic data are collected until they cover an entire set of sampling positions. Then, the kinetics are rearranged into interferograms,

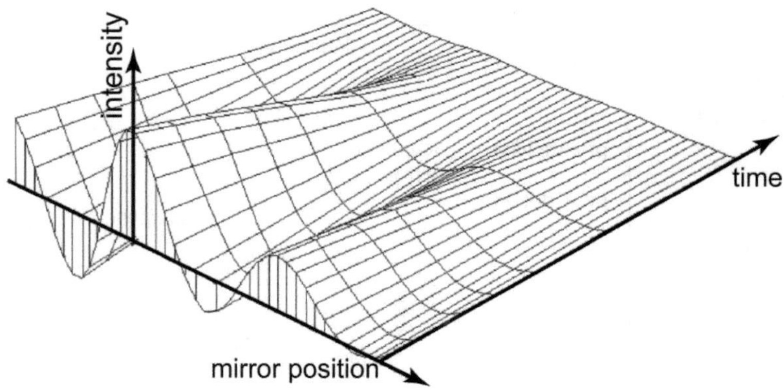

Fig. 2 Illustration of the data recording process during a step-scan FT-IR experiment. Data points are collected along the time axis successively at a given set of mirror positions, which constitute the interferogram. After the step-scan experiment, the recorded transients at the various mirror positions are rearranged into interferograms at certain times after starting the reaction of the protein

and single-channel spectra are computed (vide supra). Finally, transient absorption difference spectra are calculated in the same manner like in the rapid-scan technique.

In order to reduce the acquisition time one can take advantage of undersampling (Griffiths and de Haseth 1986). The total number of step points in the interferogram depends on the spectral resolution and bandwidth. Since for protein conformational changes there is little spectral information above 1,800 cm^{-1}, an optical low-pass filter can be used that limits the free spectral change from 1,950 to 950 cm^{-1}. Recording the spectra with 4.5 cm^{-1} resolution, the number of sampling points for a one-sided interferogram is reduced to 844.

2.2.1 Step-Scan Device

Figure 1 shows the experimental step-scan device employed by our group. It consists of a Bruker IFS 66v/S spectrometer with an MCT detector connected to a DC-coupled preamplifier, a 200 KHz, 16 bit on-board transient recorder, and an external programmable digital pulse generator (Model 39, Wavetek, Ismaning, Germany). The optical bench is evacuated down to 3 mbar and is mounted on a vibrationally isolated table. These are the prerequisites for an improved stability of the moving mirror in the step-scan mode.

For pulsed sample excitation, a Q-switched Nd:YAG laser (GCR 12S, Spectra Physics, Darmstadt, Germany, frequency-double output at 532 nm, pulse duration of 8 ns, maximum excitation energy of 100 mJ/cm^2) is used. Dielectric mirrors are used to direct the laser emission to the sample. The repetition rate of the excitation is adjusted depending on the decay time of the transient species. The pulse-to-pulse intensity varies by as much as 10% causing variations of the number of excited molecules with every repeated event. To compensate for these variations which ultimately leads to additional noise in the spectra, several laser shots are co-added at each sampling position to ensure a reproducible average value.

2.2.2 Step-Scan Timing

In a step-scan experiment, the mirror movements, laser trigger, and data acquisition have to be synchronized. A suitable method is to use an external pulse generator to master the exact sequence of the timing. The interferometer controller steps and stabilizes the mirror at a certain position and sends a signal to the on-board analogue-to-digital converter (ADC) to start data collection with the next signal from the external experiment trigger source. After an appropriate delay, this generates a TTL-pulse which is sent concomitantly to the laser flashlamps and the digitizer. The flashlamps pump the laser medium and after 140 μs the Q-switch is activated and a laser flash is emitted. Data recorded during the pre-triggering time are used as reference for difference spectroscopy. After sample excitation, the digitizer is triggered by the logarithmic clock of the external programmable digital pulse generator. In a typical experimental run, 1,000 time points cover the time range from 7 μs to 200 ms.

2.2.3 Step-Scan Parameters

With the step-scan technique, the time resolution is not restricted by the velocity of the interferometer mirror, but is limited by the response time of the detection system as well as by the digitization rate of the transient recorder. Thus, it is important that the electronic components of the step-scan device have comparable bandwidths and speeds to perform the experiment. It is desirable to fit the time resolution to the dynamics of the studied event, because the higher the time resolution the noise will increase. Taking these observations into account, Chen and Palmer (1997) reported practical realizations of step-scan FT-IR measurements with 10 ns time resolution. They characterized the decay kinetics of a transition metal complex after photodissociation of ligated CO. It is pointed out, however, that the difference absorbance observed in experiments with proteins is at least one order of magnitude less intense.

3 Applications: Bacteriorhodopsin

Numerous time-resolved infrared studies were focused on the small integral membrane protein bacteriorhodopsin (BR) which is the best-understood member of the seven-helical transmembrane protein family. When illuminated, this small (26 kDa) but very robust protein pumps protons out of the cell to establish a proton gradient across the membrane that drives ATP synthesis.

Upon light excitation, BR undergoes a series of light-induced cyclic reactions, of which the primary event is the isomerization of the all-*trans* retinal chromophore to the 13-*cis* configuration. As a consequence, BR passes throught a series of intermediate states termed J, K, L, M, N and O, which were well characterized by visible spectroscopy (Lozier et al. 1975, Chizhov et al. 1996).

One of the first applications of step-scan FT-IR spectroscopy with a time resolution of 500 ns investigated the KL-to-L transition, demonstrating that these two states are in equilibrium (Weidlich and Siebert 1993). A clear-cut separation of the subsequent L, M, N and O intermediates was achieved by time-resolved step-scan ATR/FT-IR spectroscopy by varying temperature and pH values (Zscherp and Heberle 1997, Fig. 7).

On the basis of the IR difference spectra in the carbonyl region (Fig. 3a), the assignment of the bands in the various intermediate states leads to a exemplary understanding of proton pump function of BR, which is discussed in six proton transfer steps (Fig. 3b). The pioneering work of Engelhard et al. (1985) revealed that besides the retinal Schiff base, aspartic acids play a dominant role in proton transfer within BR. IR spectroscopy is particularly useful when studying the role of acidic amino acids because the C=O stretching frequency is well-isolated from other vibrations of the protein.

As the present review puts the focus on the application of time-resolved FT-IR spectroscopy, we will not explicitely discuss the extensive work on the assignment of the difference bands but refer to excellent reviews on the topic (Maeda 1995,

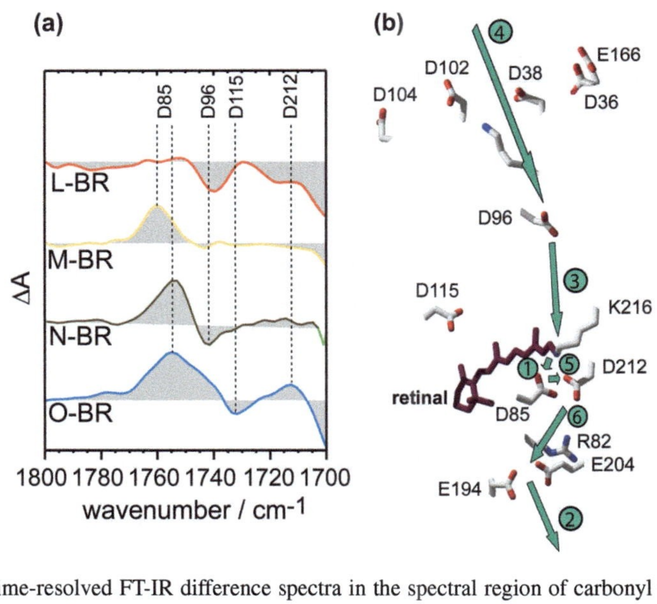

Fig. 3 **a** Time-resolved FT-IR difference spectra in the spectral region of carbonyl vibrations of bacteriorhodopsin. Early intermediate states of the photocycle are at the *top* and the late states are shown at the *bottom*. Transient protonation (*positive bands*) and deprotonation (*negative bands*) of four key aspartic amino acids are labeled which are involved in proton pumping of bacteriorhodopsin. **b** Sequence of proton transfer steps (*green arrows*) which constitute the proton pump function of bacteriorhodopsin, as derived from the difference spectra shown in (**a**)

Heberle et al. 2000, Heberle 2000, Dioumaev 2001). The negative band feature in the L-BR difference spectrum arises from the C=O stretching vibration of D96 and D115 (Fig. 3a, red trace). These amino acid residues are both protonated in the ground-state BR and during the conversion to the L intermediate they undergo changes in hydrogen bonding. The M state which forms in about 50 μs, is characterized by proton transfer from the retinal Schiff base to D85 (step 1 in Fig. 3b). A strong band at 1,761 cm^{-1} characterizes this proton transfer event which is due to the C=O stretch of the sidechain of D85 (yellow trace in Fig. 3a). The proton release reaction of an excess proton from the extracellular side of the membrane was identified by nanosecond time-resolved FT-IR spectroscopy (Garczarek et al. 2004, Garczarek and Gerwert 2006, Rammelsberg et al. 1997, 1998). The authors proposed a proton release pathway in the shape of a transient hydrogen-bonded network comprising several polar amino acids and water molecules (step 2 in Fig. 3b). Evidence is growing that such a dynamic water-assisted local area network (WLAN) (Heberle 2004, Mathias and Marx 2007) may operate also in other ion pumps (Breton and Nabedryk 1998).

It is evident from Fig. 3a that the C=O stretching vibration of D85 shifts down by about 7 cm^{-1} when the later intermediates N and O are formed. In the N state, whose maximum concentration is reached at about 3 ms after photoexcitation (step 3), a negative band is discernible at 1,742 cm^{-1} (green spectrum in Fig. 3a). This band was assigned to the C=O stretch of D96, the internal proton donor to the Schiff base. Reprotonation of D96 from the cytoplasmic surface takes place in the late ms (step 4) and is strongly pH-dependent (Zscherp et al. 1999). Protons from the cytoplasm are attracted by negatively charged residues along the surface and funneled to the entrance of the proton uptake pathway where D38 is located (Riesle et al. 1996). In the final stages of the photoreaction, D85 deprotonates and the released proton is transferred to the nearby residue D212 (step 5 in Fig. 3b) for which the C=O stretch is assigned to the positive band at 1,713 cm^{-1} (blue in Fig. 3a). Under physiological conditions the residence time for the proton at D212 is short and the proton is readily transferred to the initial proton release complex (step 6). This reaction completes the proton transfer reaction across BR. In conclusion, the electromechanical driving force for proton translocation is provided by photon absorption of the retinal and the consequent structural changes of the cofactor and the apo-protein.

The interpretation of the results of IR spectroscopy with respect to details of the reaction mechanism relies also on structural data. The elucidation of the molecular structure of BR at nearly atomic resolution (Luecke et al. 1999), however, raises the pertinent question whether the catalytic activity of the protein(s) is preserved in the microcrystals. Initially, the crystals were investigated by steady-state IR spectroscopy and the difference spectra indicated unperturbed intermediate states (Heberle et al. 1998). In a further step, we succeeded to study single BR crystals with a diameter of 50 μm by time-resolved step-scan FT-IR microspectroscopy (Efremov et al. 2006). It was found that similar structural changes occur both in the crystal and in the purple membrane though with slightly different kinetics.

4 New Developments of the Step-Scan FT-IR Technique

Time-resolved step-scan FT-IR technique is mostly applied to biological systems undergoing a rapid and reversible reaction. Depending on the size of the protein, at least 100,000 repetitions of the step-scan experiment are necessary to optimize the signal-to-noise ratio (Heitbrink et al. 2002). This condition is easily achieved in studies of light-activated proteins (Uhmann et al. 1991, Zscherp and Heberle 1997) where the sample is excited by a laser pulse, measured and recovers back to the initial state before the next reaction is induced.

However, slow-cycling systems or irreversible reactions are difficult to be investigated with the classical step-scan FT-IR approach due to the low rate of sample excitation. To overcome this obstacle, it would be necessary that in each measurement the laser pulse excites a fresh sample of identical concentration. For this purpose, a flow-flash experiment in a miniaturized flow channel was developed which was integrated into a step-scan FT-IR spectroscopic setup. Figure 4 illustrates the configuration of the microfluidic device developed by the group of B. Lendl (Kaun et al. 2006). As a proof-of-principle, we studied the rebinding reaction of CO to myoglobin after photodissociation. Under a continuous flow of protein solution, it was possible to resolve small structural changes of the protein backbone which were assigned to the ligand-free form of cytochrome c, as well as the ongoing relaxation of the protein after the ultrafast dissociation of the ligand (Schleeger et al. 2009).

The use of microfluidics reduces the sample consumption to only a few 10 μl microliters of a millimolar sample solution which makes this method particularly

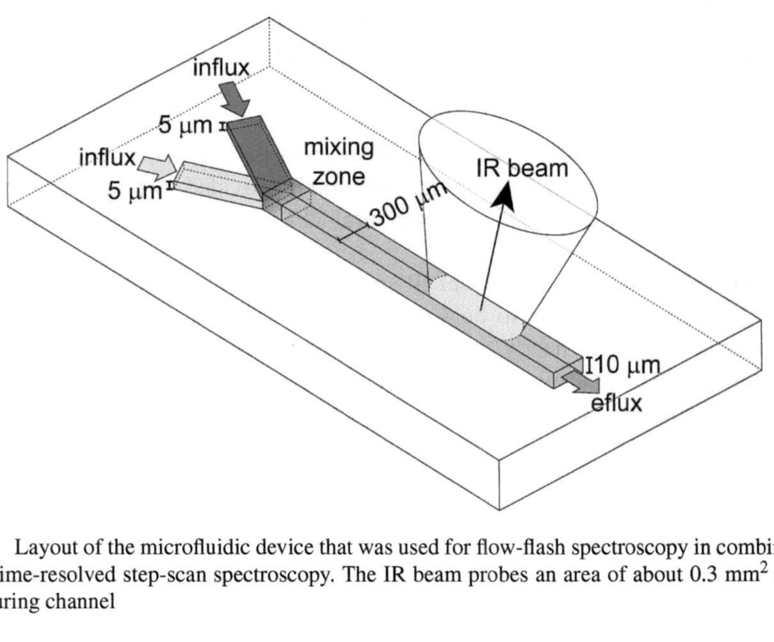

Fig. 4 Layout of the microfluidic device that was used for flow-flash spectroscopy in combination with time-resolved step-scan spectroscopy. The IR beam probes an area of about 0.3 mm^2 of the measuring channel

interesting for the investigation of expensive biological samples. Moreover, the flow cell provides the unique opportunity to assess the reaction mechanism of proteins that cycle slowly or react irreversibly. We infer that this novel approach will help in the elucidation of molecular reactions in non-cycling systems.

References

Breton, J. and Nabedryk, E. (1998) Proton uptake upon quinone reduction in bacterial reaction centers: IR signature and possible participation of a highly polarizable hydrogen bond network. *Photosynth. Res.* **55**: 301–307.

Chen, P.Y. and Palmer, R.A. (1997) Ten-nanosecond step-scan FT-IR absorption difference time-resolved spectroscopy: applications to excited states of transition metal complexes. *Appl. Spectrosc.* **51**(4): 580–583.

Chizhov, I., Chernavskii, D.S., Engelhard, M., Mueller, K.H., Zubov, B.V. and Hess, B. (1996) Spectrally silent transitions in the bacteriorhodopsin photocycle. *Biophys. J.* **71**(5): 2329–2345.

Dioumaev, A.K. (2001) Infrared methods for monitoring the protonation state of carboxylic amino acids in the photocycle of bacteriorhodopsin. *Biochemistry (Mosc)* **66**(11): 1269–1276.

Efremov, R., Gordeliy, V.I., Heberle, J. and Büldt, G. (2006) Time-resolved microspectroscopy on a single crystal of bacteriorhodopsin reveals lattice-induced differences in the photocycle kinetics. *Biophys. J.* **91**(4): 1441–1451.

Engelhard, M., Gerwert, K., Hess, B., Kreutz, W. and Siebert, F. (1985) Light-driven protonation changes of internal aspartic acids of bacteriorhodopsin: an investigation by static and time-resolved infrared difference spectroscopy using [4-13C]aspartic acid labeled purple membrane. *Biochemistry* **24**(2): 400–407.

Garczarek, F., Wang, J., El-Sayed, M.A. and Gerwert, K. (2004) The assignment of the different infrared continuum absorbance changes observed in the 3000–1800-cm^{-1} region during the bacteriorhodopsin photocycle. *Biophys. J.* **87**(4): 2676–1682.

Garczarek, F. and Gerwert, K. (2006) Functional waters in intraprotein proton transfer monitored by FTIR difference spectroscopy. *Nature* **439**(7072): 109–112.

Griffiths, P.R. and de Haseth, J.A. (1986) *Fourier transform infrared spectrometry*. Wiley, New York, NY, Vol. 83.

Heberle, J., Büldt, G., Koglin, E., Rosenbusch, J.P. and Landau, E.M. (1998) Assessing the functionality of a membrane protein in a three-dimensional crystal. *J. Mol. Biol.* **281**(4): 587–592.

Heberle, J. (2000) Proton transfer reactions across bacteriorhodopsin and along the membrane. *Biochim. Biophys. Acta* **1458**(1): 135–147.

Heberle, J., Fitter, J., Sass, H.J. and Büldt, G. (2000) Bacteriorhodopsin: the functional details of a molecular machine are being resolved. *Biophys. Chem.* **85**(2–3): 229–248.

Heberle, J. (2004) A local area network of protonated water molecules. *Biophys. J.* **87**(4): 2105–2106.

Heitbrink, D., Sigurdson, H., Bolwien, C., Brzezinski, P. and Heberle, J. (2002) Transient binding of CO to Cu_B in cytochrome c oxidase is dynamically linked to structural changes around a carboxyl group: a time-resolved step-scan Fourier transform infrared investigation. *Biophys. J.* **82**(1 Pt 1): 1–10.

Kaun, N., Kulka, S., Frank, J., Schade, U., Vellekoop, M.J., Harasek, M. and Lendl, B. (2006) Towards biochemical reaction monitoring using FT-IR synchrotron radiation. *Analyst* **131**(4): 489–494.

Lozier, R.H., Bogomolni, R.A. and Stoeckenius, W. (1975) Bacteriorhodopsin: a light-driven proton pump in Halobacterium Halobium. *Biophys. J.* **15**(9): 955–962.

Luecke, H., Schobert, B., Richter, H.T., Cartailler, J.P. and Lanyi, J.K. (1999) Structure of bacteriorhodopsin at 1.55 Å resolution. *J. Mol. Biol.* **291**(4): 899–911.

Maeda, A. (1995) Application of FTIR spectroscopy to the structural study on the function of bacteriorhodopsin. *Isr. J. Chem.* **35**: 387–400.

Manning, C.J., Palmer, R.A. and Chao, J.L. (1991) Step-scan Fourier-transform infrared spectrometer. *Rev. Sci. Instrum.* **62**(5): 1219–1229.

Mathias, G. and Marx, D. (2007) Structures and spectral signatures of protonated water networks in bacteriorhodopsin. *Proc. Natl. Acad. Sci. USA* **104**(17): 6980–6985.

Rammelsberg, R., Heßling, B., Chorongiewski, H. and Gerwert, K. (1997) Molecular reaction mechanism of proteins monitored by nanosecond step-scan FT-IR difference spectroscopy. *Appl. Spectrosc.* **51**(4): 558–562.

Rammelsberg, R., Huhn, G., Lübben, M. and Gerwert, K. (1998) Bacteriorhodopsin's intramolecular proton-release pathway consists of a hydrogen-bonded network. *Biochemistry* **37**(14): 5001–5009.

Riesle, J., Oesterhelt, D., Dencher, N.A. and Heberle, J. (1996) D38 is an essential part of the proton translocation pathway in bacteriorhodopsin. *Biochemistry* **35**(21): 6635–6643.

Schleeger, M., Wagner, C., Vellekoop, M.J., Lendl, B. and Heberle, J. (2009) Time-resolved flow-flash FT-IR difference spectroscopy: the kinetics of CO photodissociation from myoglobin revisited. *Anal. Bioanal. Chem.* **394**(7): 1869–1877.

Uhmann, W., Becker, A., Taran, C. and Siebert, F. (1991) Time-resolved FT-IR absorption spectroscopy using a step-scan interferometer. *Appl. Spectrosc.* **45**: 390–397.

Weidlich, O. and Siebert, F. (1993) Time resolved step-scan FTIR investigations of the transition from KL to L in the bacteriorhodpsin photocycle: identification of chromophore twists by assigning hydrogen-out-of-plane (HOOP) bending vibrations. *Appl. Spectrosc.* **47** (9):1394–1400.

Zscherp, C. and Heberle, J. (1997) Infrared difference spectra of the intermediates L, M, N, and O of the bacteriorhodopsin photoreaction obtained by time-resolved attenuated total reflection spectroscopy. *J. Phys. Chem. B* **101**(49): 10542–10547.

Zscherp, C., Schlesinger, R., Tittor, J., Oesterhelt, D. and Heberle, J. (1999) In situ determination of transient pKa changes of internal amino acids of bacteriorhodopsin by using time-resolved attenuated total reflection Fourier-transform infrared spectroscopy. *Proc. Natl. Acad. Sci. USA* **96**(10): 5498–5503.

Recombinant Membrane Protein Production: Past, Present and Future

Ravi K.R. Marreddy, Eric R. Geertsma, and Bert Poolman

Abstract One of the major challenges in membrane protein structural genomics is the production of properly folded protein in large quantities. Several *in cell* and cell-free expression systems have been developed. However, in most cases laborious trial-and-error based optimization of either the host, genetic circuitry or protein is necessary for high level production. A better understanding of membrane protein biogenesis is needed to obtain further insights into the bottlenecks of their expression. The application of "Omics" technologies to understand the host cell response to membrane protein overproduction has contributed significantly to our understanding of membrane protein production and provided rationales for optimization both the host cells and/or expression conditions. In this review, we present an overview of the current well-established expression systems and the successful approaches to optimize the synthesis of well-folded and functional membrane proteins.

Keywords Transport proteins · Recombinant expression · Membrane protein biogenesis · *Lactococcus lactis* · *Escherichia coli* · Membrane protein production

1 Introduction

Biological membranes form a barrier between the inside of cells or cellular organelles and their surrounding environment. Biological membranes are composed of lipids, arranged in bilayers, and proteins in typical lipid-to-protein ratios ranging from 3:1 to 1:3 (w/w) (Boon and Smith 2002). A typical plasma membrane has a lipid-to-protein ration of 1:1, which corresponds to approximately 25,000 membrane proteins per μm^2 and only a few layers of lipids per protein. The

---

B. Poolman (✉)
Department of Biochemistry, Groningen Biomolecular Sciences and Biotechnology Institute, Netherlands Proteomics Centre and Zernike Institute for Advanced Materials, University of Groningen, Nijenborgh 4, 9747 AG Groningen, The Netherlands
e-mail: b.poolman@rug.nl

J. Brnjas-Kraljević, G. Pifat-Mrzljak (eds.), *Supramolecular Structure and Function 10*, DOI 10.1007/978-94-007-0893-8_3,

membrane forms a permeability barrier and translocation of solutes and transduction of signals between the outside and the inside medium is carried out by proteins embedded in the lipid bilayer. Membrane proteins are involved in diverse cellular processes such as transport of nutrients/metabolites, sensing of environmental changes and signal transduction, energy transduction and scaffolding of the structure of the cell. For most sequenced genomes, a large fraction (20–30%) of the open reading frames is predicted to code for membrane proteins. Despite this large fraction of different membrane proteins, the total number of unique membrane protein structures deposited in the protein database (PDB) is strikingly low and less than 1% (White 2009). Progress in the structural and functional characterization of membrane proteins is mostly hindered by a generally low abundance in cells and by their amphipathic nature, which requires the use of detergents and often tolerates only careful manipulation. As only few membrane proteins are naturally abundant in a certain membrane, structural and functional analysis virtually always requires establishing an overexpression system. Regrettably, the quality and quantity of the overproduced membrane proteins are often unpredictable using the currently available expression strategies, and successful overexpression often requires extensive fine-tuning. Although in general more difficult than for water soluble proteins, overcoming the expression hurdle for membrane proteins is certainly not impossible. Over the past two decades several successful expression strategies for membrane proteins have been reported (Grisshammer 2006, Tate 2001, Tate and Grisshammer 1996), but development of these systems was mostly of empirical nature. Improved insights in the biogenesis of membrane proteins and the start of the identification of bottlenecks encountered during membrane protein overexpression (Marreddy et al. 2010, Wagner et al. 2008a), now allow a more rational approach (Wagner et al. 2006). Despite this progress, our current knowledge of the process is far from complete and even further from being predictive. Keeping this in mind, the present best approach to establish an overexpression system involves a broad sampling of "expression space". Here, we shortly describe our current understanding of membrane protein biogenesis before presenting an overview of several expression strategies and studies aimed at understanding and overcoming the bottlenecks in the functional expression of membrane proteins. We focus on recent developments in expression optimization in microorganisms but also refer to studies in (higher) eukaryotes when relevant.

2 Membrane Targeting and Insertion

Compared to most soluble proteins, the biogenesis of membrane proteins is more complex and requires machinery to target the proteins and to insert them into the membrane. Both in pro- and eu-karyotic cells, the majority of the membrane protein translocation occurs through a universally conserved Sec (secretory) pathway. The following steps are crucial for a successful translocation and integration of proteins into the membrane in functional form: (i) identification of the protein to be translocated; (ii) discrimination between the protein to be translocated into the membrane and proteins to be secreted; (iii) export and integration of the proteins into or across the lipid bilayer; and (iv) functional folding of the membrane integrated protein without disrupting the membrane integrity.

The membrane protein substrates that need to be inserted into the membrane via the Sec pathway are identified by the Signal Recognition Particle (SRP). The SRP is universally conserved and is composed of an RNA molecule and a variety of protein components (Table 1), with the chloroplast being the only exception as this SRP does not comprise a RNA molecule (Schunemann 2004). The Ribosome Nascent Chain (RNCs) emerging from the translating ribosome is recognized by

Table 1 Components of the various membrane protein biogenesis pathways

	Gram-negative	Gram-positive					
	E. coli	*B. subtilis*	*L. lactis*	Yeast	Mammals	Chloroplast	Archaea
1. Sec-mediated protein translocation							
1a. Signal recognition particle							
RNA molecule	4.5S RNA	6S RNA (scRNA)	4.5S RNA	11S RNA (scR1)	7S RNA	–	7S RNA
SRP complex	Ffh	Ffh Hbsu	Ffh	SRP14 SRP19 SRP21 SRP54 SRP68 SRP72	SRP9 SRP14 SRP19 SRP21 SRP54 SRP68 SRP72	SRP43 SRP54	SRP19 SRP54
SRP receptor	FtsY	FtsY	FtsY	SRα SRβ	SRα SRβ TRAM	cpFtsY	FtsY
1b. Sec complex							
ATPase	SecA	SecA	SecA	–	–	cpSecA	–
Integral membrane pore complex (Sec pore)	SecY SecE SecG	SecY SecE SecG	SecY SecE SecG	Sec61p Sss1p Sbh1p	Sec61α Sec61γ Sec61β	cpSecY cpSecE	SecY SecE Secβ
Sec pore associated proteins	SecD SecF YajC YidC	SecD SecF YajC	YajC	BiP	BiP TRAP complex RAMP4		SecD SecF
2. YidC mediated protein translocation							
YidC components	YidC	SpoIIIJ YqjQ	OxaA2 Llmg_04 13	OxaA1	–	Alb3	Hsp[a]

[a] Not present in all the members of the taxa, – Not present.

the SRP (Koch et al. 1999, Walter et al. 1981), and the presence or absence of a positively charged N-terminal signal sequence in the RNCs differentiates between secretory proteins and membrane proteins. The N-terminus of secretory proteins generally possess one to three positively charged amino acid residues followed by 10–15 hydrophobic residues, which act as a signal sequence (or presequences). Most preproteins, in *Escherichia coli* these are the periplasmic and outer membrane proteins, are targeted to the Sec translocase via the molecular chaperone SecB; this targeting route will not be discussed in this review. Membrane proteins in general do not possess a signal sequence, but their first hydrophobic transmembrane segment functions as a signal for membrane targeting and insertion (Huber et al. 2005, Ng et al. 1996). The SRP binds to the first hydrophobic transmembrane segment of the RNCs and then targets the protein-ribosome complex to the membrane. The nascent proteins are subsequently released into the heterotrimeric protein conducting channel, called Sec-translocon (Table 1). The transfer of a nascent polypeptide into the Sec pore requires energy and is driven by SecA-mediated ATP/GTP hydrolysis in bacteria and by trapping by *Bi*nding *P*rotein (BiP) in eukaryotes. Based on the mode of translocation, two general mechanisms for Sec-dependent protein translocation are distinguished: (i) Co-translational translocation, in which the protein translation is arrested upon binding of SRP to RNCs, proceeds once the SRP-RNC complex binds to the membrane receptor FtsY (see Fig. 1); and (ii) Post-translational translocation, in which proteins are transported to the membrane once their synthesis is complete. Integral membrane proteins of the cytoplasmic membrane use the path of co-translational translocation, whereas secreted or outer membrane proteins follow the path of post-translational translocation; the proteins are kept unfolded by specific chaperones such as SecB in *Escherichia coli*. (For detailed mechanistic insight into these two modes of protein translocation we refer to Rapoport 2007).

Besides the Sec pathway, a simpler route for membrane insertion of a subset of proteins has been identified. This pathway makes use of the evolutionarily conserved YidC, Oxa and Alb3 proteins that have been identified in bacteria, mitochondria and chloroplast, respectively (Dalbey and Kuhn 2004). Although these proteins can function in association with the Sec pathway in bacteria and chloroplasts, they have also been proposed to catalyze protein translocation independent of the Sec-system (Chen et al. 2002, Serek et al. 2004). The membrane insertion of a number of proteins has been shown to be strictly dependent on YidC (or its homologues), such as subunit *c* of the F_1F_0-ATP synthase of mitochondria and bacteria (Altamura et al. 1996, van der Laan et al. 2004) and light-harvesting chlorophyll-binding proteins in chloroplast (Moore et al. 2000). Also, YidC has been shown to play a role in membrane protein folding (Scotti et al. 2000), that is, the SecYEG-associated YidC can provide a post-insertional chaperone function (Nagamori et al. 2004, Wagner et al. 2008b). YidC is conserved in all bacteria and euryarchaea but homologues have not been found in eukaryotes and crenarchaea. Interestingly, in contrast to Gram-negative bacteria, the genomes of some Gram-positive bacteria encode two YidC proteins (YidC1 and YidC2). Deletion and complementation studies in *Streptococcus mutans* have shown that YidC1 and YidC2 are paralogues (Hasona et al. 2005). There is evidence that the function of YidC2 from *S. mutans* is similar

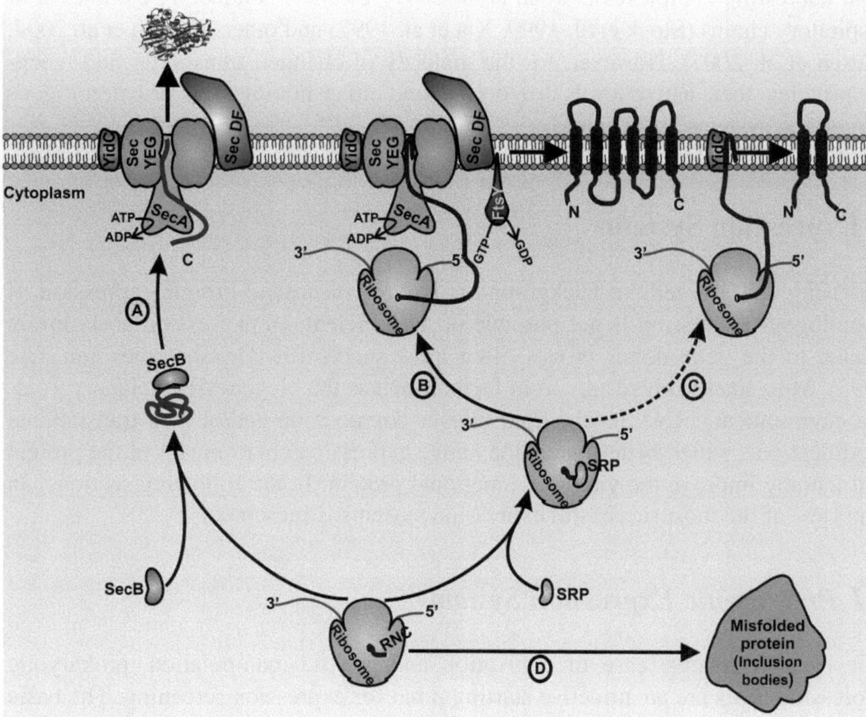

Fig. 1 Schematic representation of the paths for membrane protein biogenesis. The scheme is based on known components in *E. coli* (protein chaperones other than SecB, foldases and membrane-bound proteases are not indicated) (*A*) Sec-mediated post-translational targeting, (*B*) Sec-mediated co-translational targeting, and (*C*) Sec-independent YidC-mediated targeting. (*D*) Misfolded proteins can end up in inclusion bodies. Abbreviations: RNC, Ribonucease nascent chain; SRP, Signal recognition particle

to that of OxaA1 from mitochondria and that this protein facilitates co-translational insertion of membrane proteins (Funes et al. 2009). The primary function of YidC1 is not yet known, but this protein is suggested to play a limited role in protein translocation as its genomic deletion in *S. mutans* is less detrimental when compared to the deletion of YidC2 (Hasona et al. 2005).

3 Naturally Abundant Membrane Proteins

Occasionally, expression levels of membrane proteins in natural tissues are sufficiently high to allow their purification and subsequent structural characterization. To date, membrane proteins highly abundant in natural sources of which the structures have been solved include porins (de Groot et al. 2001, Murata et al. 2000), proteins involved in photosynthetic reactions (Luecke et al. 1998, Palczewski et al. 2000) and light harvesting complexes (Jordan et al. 2001), electron-conducting complexes of respiratory chains (Stock et al. 1999, Xia et al. 1997) and others (Morth et al. 2007, Olesen et al. 2007). However, for the majority of channel, transporter and receptor proteins, their native levels are too low and either homologous or heterologous expression systems are required.

4 Expression Systems

Choosing the best cellular background is key for successful protein expression. If homologous expression is not possible or not efficient, an expression host closely related to the gene donor proves often most successful (Grisshammer and Tate 1995). Most likely, this results from factors such as the biogenesis machinery, folding environment, tRNA levels, lipid bilayer composition and/or post-translational modifications, which better match the native expression environment of the protein and thereby improve the yields of functional protein. In the following sections, an overview of the most successful expression systems is presented.

4.1 Prokaryotic Expression Systems

Due to their relative ease of cultivation and genetic manipulation, prokaryotic expression hosts are an attractive starting point for expression screening. The basic advantage of bacterial cells is their short generation time which offers rapid testing and optimization of the expression levels and rapid production of sufficient cell mass. Often the genetics are well described (e.g., as for *E. coli, Bacillus subtilis, Lactocococcus lactis*), and cloning vectors and mutant host strains are available or can be generated relatively easily. Despite these advantages, the expression of eukaryotic membrane proteins is often challenging in bacterial hosts, presumably due to a mismatch of membrane protein biogenesis machinery or kinetics of protein synthesis and folding. The overexpressed membrane protein can be mistargeted and/or misfolded, for instance as a result of higher rate of synthesis in bacteria than in the native eukaryotic hosts (Grisshammer 2006). Another disadvantage of bacterial cells is their inability to perform many post-translational modifications such as glycosylations. In addition, the difference in the lipid composition (bacterial cell membranes generally lack steroids, sphingolipids and polyunsaturated lipids) may compromise the activity once the protein has been inserted and assembled correctly into the membrane. Despite these caveats, the ease of working with prokaryotic hosts and the possibility to tune expression factors makes them even for membrane proteins of eukaryotic origin an option that is worthwhile exploring.

4.1.1 Escherichia Coli

The unsurpassed ease of genetic manipulation, the wealth of documented case studies and fast generation time (doubling of 20–60 min, depending on the growth medium) make the Gram-negative bacterium *E. coli* generally the host of first choice for initial screening of recombinant expression. Many tools have been developed for genome engineering and gene cloning and expression. Both complex broths and chemically defined media are available and cells can be grown to very high densities (OD_{660} of 10–100) at an optimal temperature of 37°C. A wide variety of plasmids with copy numbers ranging from ~2 up to >100 per cells is available. Inducible promoters with various strength ranging from relatively weak (*lac*) to moderately high (*trc*, *tac*, *tetA*, *araP$_{BAD}$* and *rhaP$_{BAD}$*) and very strong (T7 and *lac*/T5) are available (Hannig and Makrides 1998, Terpe 2006). Importantly, various plasmid systems that facilitate high-throughput cloning and expression trials are available, such as Gateway^{TM}, Ligation-independent cloning, and the Univector plasmid-fusion system (Aslanidis and Jong 1990, Liu et al. 1998, Walhout et al. 2000). Several specialized strains have been developed to overcome problems commonly observed when overexpressing proteins, e.g., the Rosetta strains (Novagen) with enhanced levels of tRNAs for rare codons, strains lacking a specific protease such as BL23(DE3) (devoid of OmpT and Ion proteases), strains with enhanced disulfide bond formation, e.g. the Origami strains from Novagen, and others. In addition, several approaches are available to co-express factors whose low-abundance forms a bottleneck limiting the correct folding of proteins (Chen et al. 2003, de Marco 2007, Gonzalez-Montalban et al. 2007, Link et al. 2008).

Membrane proteins can be expressed in *E. coli* either in a correctly-folded and membrane-inserted state or as insoluble cytosolic aggregates (inclusion bodies). These inclusion bodies can be dissolved and sometimes the proteins can be refolded to their native state (Bannwarth and Schulz 2003, Kiefer 2003). This procedure is relatively efficient for outer membrane β-barrel proteins (Liang and Tamm 2007). Protein production in inclusion bodies holds a number of advantages over expression of well-folded material: (i) it is often less a burden to the cell; (ii) inclusion bodies are less prone for degradation; (iii) toxic proteins are mostly non-toxic in inclusion bodies; (iv) the target protein can be isolated relatively pure from inclusion bodies; and (v) high expression levels can be reached. Despite these advantages, the poor success rates of the renaturation procedure for α-helical membrane proteins makes expression in their functional membrane-inserted state the preferred option.

Though not completely controlled, it is often possible to influence the state in which the membrane protein is overexpressed, that is, either functionally in the membrane or as soluble aggregate. Inclusion bodies mostly emerge when the transcription and translation rates are high, such as when a fully-induced and strong promoter is used in combination with a growth temperature that supports fast growth. To avoid inclusion bodies, one can use tunable promoters of intermediate strength, media that allow a more gradual onset of expression (Gordon et al. 2008) or a slower growth at temperatures significantly below the one supporting optimal growth (Geertsma et al. 2008, Quick and Wright 2002, Wagner et al. 2008).

4.1.2 Lactococcus Lactis

Over the past decade, the Gram-positive bacterium *Lactococcus lactis* has emerged as an alternative and complementary host for heterologous protein production (Kunji et al. 2003, Monne et al. 2005, Mulligan et al. 2009, Niu et al. 2008, Quick and

Javitch 2007). *L. lactis* is a facultative anaerobe that grows rapidly in complex or synthetic media (doubling times of 40–50 min) and proceeds to intermediate cell densities (OD_{600} ~5) at an optimal temperature of 30°C. Various constitutive (van der Vossen et al. 1987) and inducible (de Vos 1999, Llull and Poquet 2004, Miyoshi et al. 2004) promoter systems are available of which the well-tunable nisin system is used most often (de Ruyter et al. 1996, Geertsma and Poolman 2007). Compared to *E. coli,* the genetic manipulation in *L. lactis* is less efficient but high-throughput cloning-expression systems have been developed also for this host (Frelet-Barrand et al. 2010, Geertsma and Poolman 2007).

L. lactis has limited proteolytic activity (Fig. 2) and contains a single membrane with a high fraction of glycolipids. The thick cell wall requires harsher conditions

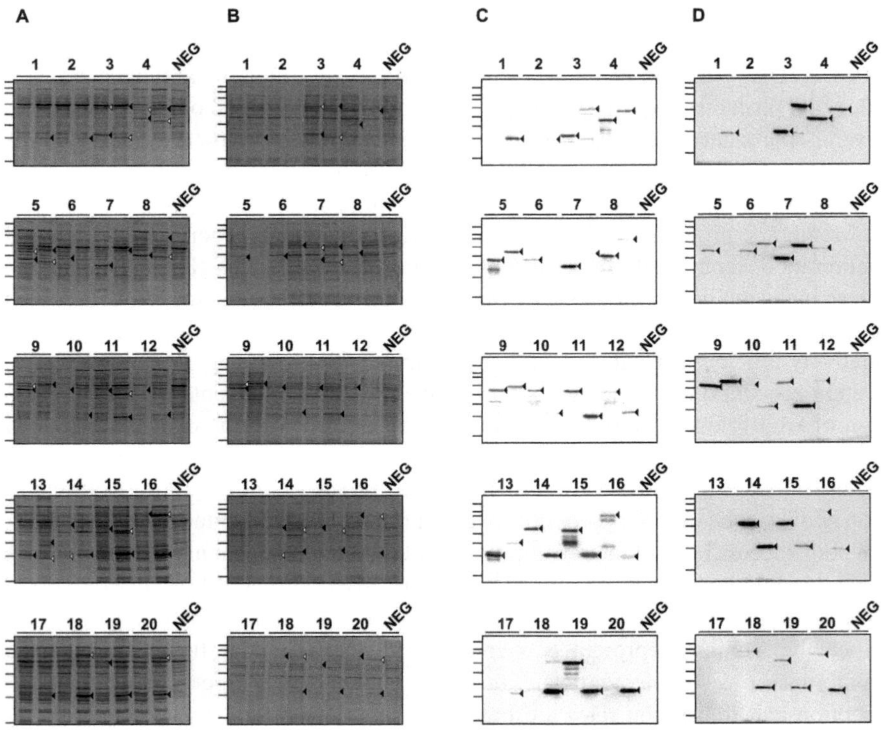

Fig. 2 *A* comparison of the expression levels of ABC transporters from the PAO and OTCN family in *E. coli* and *L. lactis* (for details of the proteins see Table 2). Membrane vesicles of *E. coli* cells (**Panels a** and **c**) or *L. lactis* (**Panels b** and **d**) overexpressing ABC transporters were submitted to SDS-PAGE and Coomassie-staining (**Panels a** and **b**) or immunoblotting with an antibody directed against the His-tag (**Panels c** and **D**). *Numbers* above the panels indicate the overexpressed ABC transporters. For each transporter, an N- (*left lane*) and C-terminal (*right lane*) His-tag was used, resulting in tagging either one of the subunits (the membrane domain or the nucleotide-binding domain). *Black* and *white arrows* indicate His-tagged and non-tagged subunits, respectively. Marker bands of 170, 130, 100, 70, 55, 40, 35, and 25 kDa are indicated on the *left*. For each gel, the most *right lane* shows a sample of membrane vesicles of cells containing an empty plasmid (labeled "NEG")

during cell disruption than needed for *E. coli,* but if this is taken into account, membrane vesicles can be prepared with high efficiencies. The limited number of endogenous transport functions present facilitates complementation studies. Being a multiple amino acid auxotroph with well-characterized transport systems for amino acids and peptides, *L. lactis* can readily be employed for incorporating amino acid analogues (Berntsson et al. 2009, El et al. 2008). In addition, the activity of the over-expressed protein can be evaluated by whole cell transport or binding assays more readily than in Gram-negative bacteria because of the absence of an outer membrane (Kunji et al. 2003, Monne et al. 2005).

The most predominantly used lactococcal strain for protein expression is the NZ9000 strain, which is a derivative of MG1363 in which the *nisRK* genes has been inserted into the *pepN* locus. *NisR* and *nisK* encode a two-component regulatory system, NisR being the transcription factor and NisK the nisin sensor, that allows controlled transcription from the *nisin A* promoter. The *L. lactis nisin A*-inducible promoter system is very tight and gene expression can be achieved in a dynamic range of more than 1,000-fold. Most recently, a strategy was devised to isolate derivatives of NZ9000 with increased functional expression of (membrane) proteins. The evolved strains have been sequenced and represent a next generation of expression hosts (Linares et al. 2010). Additionally, tools have been developed for co-expression of factors that enhance the functional overexpression of membrane proteins (Marreddy et al. 2010, Pinto et al. 2011).

Despite the fact that many pro- and eukaryotic membrane proteins have been successfully overexpressed in *L. lactis*, a direct comparison of the expression potential of *E. coli* and *L. lactis*, based on a well-balanced large set of genes from both closely related and more distant species, is lacking. A comparison of expression of 14 transport proteins from *Salmonella typhimurium* (9 targets; like *E. coli, S. typhimurium* is a member of the family of *Enterobacteriaceae*), *Aquifex aeolicus* (4 targets) and *Pyrococcus furiosus* (1 target), showed that most (5/9) of the *S. typhimurium* targets expressed equally well in both hosts and 3/9 performed better in *E. coli* and 1/9 in *L. lactis*. Expression of the genes derived from the hyperthermophilic hosts exhibited a more distinct host preference: approximately half of the target set was expressed better in either one of the hosts and the overlap was minimal (Surade et al. 2006). A similar experiment, but with a target set more focused towards Gram-positive bacteria (16 out 20 protein complexes were derived from Gram-positive gene donors (Table. 2)) was conducted by Geertsma and Schuurman-Wolters (Fig. 2). Here, a large (70%) fraction of proteins was found to express in both hosts. Approximately 10 and 20% of the target set expressed only in *L. lactis* or *E. coli*, respectively. Of the proteins produced in both hosts, 40% expressed to higher levels in *E. coli* and 30% to higher levels in *L. lactis*. Notably, for 45% of the proteins expressed in *E. coli* additional bands of lower molecular weight were observed, indicative of breakdown products, whereas this was only 10% for *L. lactis*.

Expression of the SiaQM transporter from *Haemophilus influenza* proved better in *L. lactis*; not only was more protein produced and was the breakdown significantly less, expression of the protein hardly influenced cell proliferation, whereas it completely halted the growth of *E. coli* (Mulligan et al. 2009). The sodium/tyrosine

Table 2 List of proteins tested for expression in *E. coli* and *L. lactis*

ID	Species	Protein name	Function	Subunit 1	Subunit 2
1	*Helicobacter pylori* J99	Jhp_0757 + Jhp_0758	Putative osmoregulatory transport	81790391	81859461
2	*Helicobacter pylori* 26695	HP_0818 + HP_0819	Putative osmoregulatory transport	81668990	81815217
3	*Streptococcus pneumoniae*	ProV + ProWX	Proline-choline transport	81845044	81744088
4	*Listeria monocytogenes* EGD-e	Lmo1421 + Lmo1422	Putative osmoregulatory transport	16410850	16803462
5	*Listeria innocua*	Lin1460 + Lin1461	Putative osmoregulatory transport	16413933	81774286
6	*Closteridium acetobutylicum*	CA_C2850 + CA_C2849	Proline/glycine betaine transport	81854773	81775440
7	*Lactococcus lactis* subsp. Lactis IL1403	ChoQ + ChoS	Choline transport	81856612	81783126
8	*Streptomyces coelicolor* A3(2)	OpuAA +OpuABC	Glycine betaine transport	6119663	6119662
9	*Lactococcus lactis* subsp. Lactis IL1403	BusAA + BusABC	Glycine betaine transport	12724443	12724442
10	*Staphylococcus aureus* MRSA252	SAR1949 + GlnQ	Glutamine transport	49242229	49242228
11	*Listeria monocytogenes* EGD-e	Lmo2250 + Lmo2251	Putative amino acid transport	16804289	16804290
12	*Listeria monocytogenes* EGD-e	Lmo0847 + Lmo0848	Putative amino acid transport	16802888	16802889
13	*Anabena variabilis* ATCC 29413	Ava_0988 + Ava_0988	Putative amino acid transport	75700936	75700937
14	*Listeria innocua*	Lin2352 + Lin2353	Putative amino acid transport	81525616	81853884
15	*Listeria innocua*	Lin0840 + Lin0841	Putative amino acid transport	81527273	81854009
16	*Streptococcus agalactiae* NEM316	GlnP + GlnQ	Glutamine transport	25011575	25011576
17	*Clostridium acetobutylicum*	CA_C0111 + CA_C0112	Glutamine transport	81531334	81854941
18	*Streptococcus pneumoniae* TIGR4	SP_1241 + SP_1242	Amino acid transport	81531889	81855021
19	*Streptococcus pneumoniae* TIGR4	SP_0453 + SP_0452	Amino acid transport	15900370	81855087
20	*Lactococcus lactis* subsp. Lactis IL1403	GlnP + GlnQ	Glutamine transport	81537584	81856547

ID: the number corresponds to protein number in SDS-PAGE gels shown in Fig. 2.
Subunit: the number corresponds to protein accesion number in NCBI database.

transporter Tyt1 from *Fusobaterium nucleatum* was expressed in both hosts, but the fraction of functional protein was higher in *L. lactis* (Quick and Javitch 2007). Improved functional expression levels in *L. lactis* were also observed for mitochondrial transport carriers and the human KDEL-receptor (Kunji et al. 2003). The development of *L. lactis* as an alternative and complementary expression host for membrane proteins started relatively late as compared to *E. coli*. Despite the shorter timeframe, the lactococcal expression suite is very complete and easy to use, though some areas, such as high-density fermentation, require further developments.

4.1.3 Alternative Bacterial Expression Systems

The Gram-positive bacterium *Bacillus subtilis* is another alternative host for recombinant membrane protein expression. The relative ease of genetic manipulation, fast generation times and the availability of various expression systems prompted the expression trials for various membrane proteins (Bongers et al. 2005, Geissendorfer and Hillen 1990, Schumann 2007, Thuy Le and Schumann 2007, Zweers et al. 2009). Besides these advantages, the exploitation of this host for expression of membrane proteins is limited, in part owing to the relatively high proteolytic activity of *B. subtilis*, which manifests itself when the cells are disrupted for membrane or protein isolation. A number of groups are exploiting the unique physiology of *Rhodobacter sphaeroides* and R. capsulatus for the heterologous expression of membrane proteins (http://www.bio.anl.gov/structural_biology/membranep1.html). These non-sulphur purple bacteria offer attractive characteristics for the expression of membrane proteins with complex redox co-factors. Moreover, under growth conditions of low oxygen tension and exposure to low light, the organism produces extra membrane space (invaginated cytoplasmic membranes), which allows to accommodate more membrane proteins. Some redox enzymes that failed to be produced in *E. coli* were successfully overexpressed in functional form in *R. capsulatus* (de Smet et al. 2001, Kappler and McEwan 2002). The archaeon *Halobacterium salinarum* has been successfully used for recombinant expression of rhodopsin and derivatives (Bartus et al. 2003, Turner et al. 1999), but may have potential for membrane proteins from halophiles in general.

4.2 Eukaryotic Expression Systems

Eukaryotic expression hosts have in many cases proven to be successful for the production membrane proteins from higher eukaryotes, e.g. G-protein coupled receptors. Their ability to perform many of the posttranslational modifications, including efficient disulfide bond formation and glycosylations, offers a distinct advantage over prokaryotic expression systems. In some cases it has been shown that a particular lipid requirement, e.g. cholesterol, sphingolipids or glycolipids, can make a difference in the success of recombinant expression (Opekarova and Tanner 2003). In general, prokaryotes do not posses steroids and the lipid composition may be more symmetrical between the inner and outer leaflet than is the case in eukarya.

Although for many (if not most) membrane proteins the lipids are more than a solvent for embedding of the protein, the role of lipid composition in the functional expression of integral membrane proteins has been hardly studied.

Despite these advantages, the major limitation of higher eukaryotic expression systems is the more complicated, more costly fermentation and long generation times (16–24 h) of for instance insect and mammalian cells. Lower eukaryotes like yeasts and fungi offer the advantage of relative ease of fermentation and various post-translational modifications that mimic those of higher eukaryotes.

4.2.1 Yeasts

Of all the available eukaryotic expression hosts, unicellular yeasts are most often used for recombinant membrane protein production. Yeasts combine attractive prokaryotic features, such as relatively short generation times (doubling time ~2 h) and inexpensive cultivation, with a eukaryotic membrane protein biogenesis machinery and the ability to perform several post-translation modifications. Yeast cells are generally grown in liquid broths with hexose-sugars or alcohols or organic acids as carbon source. Various growth media are available for yeast cultivation of which the commercially available potato dextrose or yeast peptone dextrose or a chemically defined medium are commonly used. The growth of yeast requires abundant oxygen supply with an optimal temperature between 20 and 30°C and neutral pH. Yeast is well suited for membrane protein expression due to the relative ease of genetic manipulation, efficient DNA transformation, and cloning. The lipids of yeast plasma membranes are composed of phospholipids, sterols (ergosterol) and sphingolipids (inositol) (Opekarova and Tanner 2003). Numerous plasmids for recombinant membrane protein expression are available and are maintained either in episomal form or integrated in the genome.

Several yeasts have been developed as expression hosts, to name a few: *Hansenula polymorpha*, *Yarrowia lipolitica* and *Klyveromyces lactis*. Here, we briefly discus the yeast most commonly used for membrane protein overexpression: *Saccharomyces cerevisiae, Schizosaccharomyces pombe* and *Pichia pastoris*. *S. cerevisiae* has been most widely studied for the expression of membrane proteins from eukaryotic origin (Jidenko et al. 2005, Palanivelu et al. 2006). Various constitutive (PGK1) and inducible (GAL1 or GAL10) promoters are employed to fine tune the expression (Osterberg et al. 2006, White et al. 2007). In addition, various fusion tags are available that assists in rapid screening of membrane protein expression in a high-throughput fashion (Drew et al. 2008, Li et al. 2009, Newstead et al. 2007). A concern of *S. cerevisiae* is the occurrence of hyperglycosylation, which can affect the stability, folding and/or targeting of the protein (Bill 2001, Helenius and Aebi 2004). Compared to *S. cerevisiae*, *P. pastoris* may offer some advantages like (i) availability of various constitutive (GAP) and inducible (methanol-inducible AOX1) promoters (Hollenberg and Gellissen 1997, Sreekrishna et al. 1997); (ii) efficient growth of these cells in defined medium enables the incorporation of isotopes. An interesting characteristic of *S. pombe* is its mammalian-like glycosylation, which makes it attractive for mammalian membrane protein expression (Takegawa et al. 2009).

4.2.2 Insect Cells

Insect cells such as the *Spodoptera frugiperda* ovarian cell lines *Sf9* and *Sf21* are frequently used for the production of membrane proteins from higher eukaryotes, because their membrane lipid composition and protein processing machinery resembles that of mammalian cells. Insect cells are preferred over mammalian cell lines as their maintenance is relatively simpler. The expression is based on the infection of insect cells with a recombinant baculovirus containing the gene of interest behind a viral promoter. Transfected viral genomic DNA will hijack the protein synthesis machinery of the cell, leading to expression of the target gene (Luckow 1993).

Insect cells are grown in monolayers as well as in suspensions in commercially available serum free or protein free growth media. They grow at an optimal temperature between 25 and 27°C with high oxygen consumption. Unlike mammalian cells they do not require CO_2 for growth. Large-scale culturing (up to ~100 L) of insect cells with unaltered growth kinetics and expression levels has been made possible with the introduction of airlift or stirred tank reactors. The more expensive cultivation and long generation times (18–24 h), compared to bacterial and yeast systems, make the insect cell expression system less amenable for high-throughput screening. Unlike bacteria, insect cell membranes contain sterols and even cholesterol but in quantities much lower than in mammalian cells (Opekarova and Tanner 2003, Trometer and Falson 2010). Various expression systems are available of which the vectors derived from *Autographa californica* multiple nuclear polyhedrosis virus (AcMNPV) are most commonly used (Berger et al. 2004, Davies 1994); in special cases the *Bombyx mori* nuclear polyhedrosis virus (BmNPV) is used (Du et al. 2009). Major drawbacks of the viral vectors used to be their time-consuming construction (2–3 weeks) and the getting of a high enough titer for infection. The former has been simplified by the introduction of shuttle bacmids holding recombination sites that facilitate the insertion of a plasmid vector containing the gene of interest (marketed as the Bac-to-Bac system by Invitrogen) (Kost et al. 2005, Luckow et al. 1993). The expression of foreign genes is often done under the control of a strong viral promoter (e.g. Polyhedron or p10). The co-expression of various protein modifying enzymes (e.g. chaperones) to improve functional protein production has been explored in a number of cases (Summers 2006).

Several insect cells are available for recombinant protein expression of which the *Spodoptera frugiperda* ovarian cell lines *Sf9* and *Sf21*, and the High-Five cells from egg cells of *Trichoplusia ni* and *Mammestra brassicae* (Mb) are most commonly used. Various protocols for usage and optimization of cell lines, growth media, cell densities for viral infection (time of infection), and the time required for harvesting following infection of the cells are available (Elias et al. 2007, Harwood 2007). Even though insect and mammalian cells glycosylate proteins at the same sequence (Asn-X-Ser/Thr), the recombinant expression of mammalian proteins in insect cells often shows underglycosylation (Trometer and Falson 2010, Voss et al. 1993, Walravens et al. 1996). Despite these limitations, insect cells have been successfully used to overproduce GPCR's (Akermoun et al. 2005, Klaassen and DeGrip 2000), aquaporins (Hiroaki et al. 2006) and various membrane transport proteins (Gonzales et al. 2009, Jin et al. 2009, Sobolevsky et al. 2009).

4.2.3 Mammalian Cells

Mammalian cells are seemingly the ultimate choice for the recombinant expression of membrane proteins from higher eukaryotes like mammals as they provide the most native-like cellular environment. Although the expression levels are often sufficient for functional analyses (e.g. ligand-binding), the protein yields often fall short in the demands of structural studies (Mancia and Hendrickson 2007). Production of recombinant protein in mammalian cells can be accomplished either through transient transfection, viral infection, or through integration of foreign DNA into the host cellular genome. Even though the transient expression results in cell death after some days, the generation of these cell lines is fast and efficient. However with respect to the generation of stable cell lines, one needs to analyze several clones for protein expression after transfection. The protein yields variy between the transformants either due to differences in the copy number of the target gene in the genome or the shielding of the target gene by the chromatin.

The growth of mammalian cells demands for the presence of various growth factors that are derived from animal blood or calf serum. The culture and maintenance of these cell lines can be troublesome as the growth factors supplemented to growth media are prone to viral or prion contaminations. Cells are grown in suspension or as adherent cultures at an optimal temperature of 37°C in the presence of 5% CO_2. Similar to insect cells, the cost-effective maintenance and long generation times (~24 h) limits their use for large-scale recombinant protein expression. Various viral vectors are available of which the Semliki Forest Virus (SFV) is most commonly used. The recombinant expression of protein is predominantly under the control of constitutive promoters (e.g. 11 K or P10). However, if the target gene expression is toxic to the cells inducible promoters (e.g. metallothionein or glucocorticoid responsive elements) are available that can be used to tune the protein expression level to prevent detrimental effects.

A variety of cell lines are available for recombinant protein expression of which monkey kidney (COS), Chinese hamster ovary (CHO), baby hamster kidney (BHK-21), human embryonic kidney (HEK-293), human epitheloid carcinoma (HeLa) and GH3 cell lines are often used. A number of GPCR's (Felder et al. 1992, Jarvie et al. 1993, Reuben et al. 1994, Ulrich et al. 1993, Weinshank et al. 1992), the GABA transporter (Keynan et al. 1992), GLUT transporter (Schurmann et al. 1992) are expressed in levels sufficient for functional but not for structural analysis. In a few cases, protein production yields were achieved that allowed crystallographic studies (Kawate et al. 2009, Sokolova et al. 2001, Unger et al. 1999).

4.2.4 Alternative Expression Systems

Alternative to the heretofore described hosts, a number of other expression systems have been developed of which the *Leishmania tarantole,* a trypanosomatid protozoan parasite, is relatively frequently used. The production of membrane proteins, especially those with mammalian-like N-glycosylation patterns, has been successful in *L. tarantole* (Breitling et al. 2002). The basic advantage of *L. tarantole* with respect to mammalian cells is the lower cost and faster growth (doubling time of 6–8 h). Various constitutive and regulated promoter systems are available (Breitling et al. 2002, Kushnir et al. 2005). Till date several membrane receptors have been functionally overexpressed in *L. tarantole*, however their usefulness in structural genomics studies has not been reported yet (Basile and Peticca 2009).

Next to the above mentioned expression systems, the production of membrane proteins, biopharmaceuticals and glycoproteins in genetically-modified plant cells is receiving more attention. The basic advantage of plant cells is its (i) safe and inexpensive growth; (ii) low risk of contamination with pathogens and endotoxins; and (iii) capacity for post-translational modifications. Various foreign membrane proteins have been expressed in plant cells, but with a few exceptions the expression levels are low (~0.01–0.3% of total cell protein) (Daniell et al. 2001, Gomez et al. 1998, Haq et al. 1995, Mason et al. 1992, Tackaberry et al. 1999). The initial plant-based expression systems did not express human glycoproteins, which was due to the difference in the pattern of protein N-glycosylation (Gomord and Faye 2004). However, plant cells with humanized glycosylation machinery have been successfully generated (Schahs et al. 2007). The major disadvantage of this system is proteolytic degradation and gene silencing, which occurs during nuclear transformation and down-regulates the expression of recombinant proteins (Streatfield 2007). Some of the drawbacks of expression in plant cells were overcome by introducing the recombinant DNA into chloroplasts (so-called transplastomic plants) (Daniell et al. 2002, Maliga 2002). The basic advantages of chloroplast transformations are: (i) a higher copy number of recombinant genes because of the many chloroplasts in a typical photosynthetic cell; (ii) the absence of gene silencing; (iii) the possibility to express multiple genes in a single operon; and (iv) a limited toxicity for the plant cell due to protein production in the chloroplast. Many therapeutic proteins have been successfully expressed, using the chloroplast transformation technology (Chebolu and Daniell 2009), but so far there are no examples of membrane proteins expressed in this system. So, the potential is there but it has not been explored.

4.3 Cell-Free Expression

Over the past decade, cell free expression systems have developed as a potential alternative for the production of (membrane) proteins for structural and functional studies. In cell-free expression, the transcription and translation machinery derived from bacterial (*E. coli*) or eukaryotic (wheat germ, rabbit reticulocytes or insect cells) cells is supplied with the gene(s) of interest, amino acids (or amino acid analogues), and nucleotides (Kigawa et al. 2004, Madin et al. 2000, Sawasaki et al. 2005). The essential metabolites for transcription and translation are supplied throughout the reaction by either of the following two methods: (i) a continuous-exchange cell-free (CECF) method, in which a passive exchange of substrates and by- products takes place through dialysis (Kim and Choi 1996): or (ii) a two layer diffusion system in which the reaction mixture is carefully overlaid by the feed mixture without separation by a dialysis membrane (Sawasaki et al. 2002). Membrane proteins can either be produced as precipitants/aggregates or in a solubilized state in the presence of detergents. The proteins thus produced can be achieved in functional form by solubilization in detergent and subsequent reconstitution. Practical aspects of this approach have been well described (Schwarz et al. 2007, Torizawa et al. 2004). An alternative approach to this is the co-translational insertion of membrane proteins into Sec or YidC-containing proteoliposomes, which has proven successful but the yields have been low so far (Kalmbach et al. 2007, Nomura et al. 2008, Nozawa et al. 2007).

A major advantage of cell-free synthesis is the decoupling of cell viability and growth from protein expression. Furthermore, the reaction mixture can be easily manipulated by the addition of folding catalysts or lipids at any time. So far, the majority of membrane proteins produced in cell free systems were in a non-native state and required subsequent refolding and reconstitution steps. In a limited number of cases, however, functional protein was obtained in quantities amenable for structural studies (Chen et al. 2007, Li et al. 2006). A direct comparison of the expression of 120 different *E. coli* membrane proteins in a conventional *E. coli* in vivo system and the *E. coli*-derived cell-free expression system showed that 63% of proteins were expressed in the cell-free system, whereas only 44% were expressed in vivo. Qualitative assesment of the size and monodispersity of five proteins well expressed in the cell-free system showed that four of them were homogeneously produced (Savage et al. 2007).

5 Expression Tools

After choosing the appropriate production host, one may need to examine and optimize the expression conditions. The most relevant parameters will be described below:

(1) mRNA concentration. It is often observed that an increase in transcription (stronger promoter) does not lead to an increase in the yield of functional membrane protein; in fact, the opposite, a reduction in protein yield, may be observed (Geertsma et al. 2008, Marreddy et al. 2010, Shukla et al. 2007, Weiss et al. 1995, 1998). The mRNA levels depend on various factors like plasmid copy number, promoter strength and stability of the transcript. Most of these parameters can be manipulated experimentally.

Plasmid copy number influences the mRNA levels. In our hands, a low or moderate copy number is often preferred for expression in prokaryotic hosts. A high copy number of the plasmid may become a burden to the cell and negatively affect expression through growth inhibition. An ideal promoter should be tight and have a large dynamic range of expression, e.g. like the nisin A system of *L. lactis* (de Ruyter et al. 1996). As the optimal transcription level needs to be determined experimentally, inducible promoters are preferred over constitutive ones (Geertsma et al. 2008, Gordon et al. 2008, Lenoir et al. 2002). mRNA stability is an important factor in determining the endogenous levels of proteins in a cell, but the stability of the recombinant transcript has hardly ever been studied in the context of optimizing (membrane) protein expression. The presence of a transcription termination signal immediately downstream of the coding sequence of the target gene can stabilize the mRNA and thus improve the expression (Jacobson and Peltz 1996, Newbury et al. 1987).

(2) Quantitative and qualitative assessment of membrane protein expression. There are major concerns with regard to recombinant membrane protein expression: (i) the yield of protein (quantity), the integration into the membrane and the assembly of the protein in a functional form (quality). To asses the quality and quantity, one needs a fast and reliable method to detect the total expression level and preferably also the quality of the produced protein. The levels of membrane proteins are often too low to be identified by Coomassie- or silver-staining of SDS-PAGE gels. The availability of an appropriate antibody directed against the protein or specific tag engineered at the N- or C-terminus can alleviate this problem. However, the quantitative assessment of membrane protein expression by immunoblots is not always efficient as the hydrophobic nature of the proteins limits their transfer from the gel to the blot membrane (Korepanova et al. 2005). The functionality of the protein can be assessed by binding assays or enzyme/transport assays, however, their application is often not compatible with high-throughput screening of multiple expression conditions. The functionality of the protein can be assessed by fusing reporters like GFP to the C-terminus of the protein (Drew et al. 2002, 2006, Newstead et al. 2007). Several groups, including ours, have shown that fusing GFP to the C-terminus of a (membrane) protein can be used as an indicator for gross folding. If the target protein, fused N-terminal to GFP, becomes misfolded during the biosynthesis, it drags GFP in a misfolded, SDS-sensitive state. If the protein becomes properly folded the GFP barrel will be synthesized as a fluorescent SDS-resistant moiety; the SDS-sensitive and SDS-resistant conformations can be readily discriminated on SDS-PAGE and immunoblots (Drew et al. 2001, Geertsma et al. 2008). GFP fusions also allow to screen for cells with high-level expression by fluorescence-activated cell sorting (FACS) (Link et al. 2008, Mancia et al. 2004, Niebauer et al. 2004).

6 Host Screening/Strain Selection

Miroux and colleagues have developed variants of *E. coli* BL21(DE3), named C41(DE3) and C43(DE3), with improved membrane protein production capabilities. The selection of these strains was based on their ability to survive the toxicity imposed by the overexpression of a membrane protein in inclusion bodies. Strains C41(DE3) and C43(DE3) are in fact first and second generation mutants that grow to higher density when (over)expressing membrane proteins (Miroux and Walker 1996). The expression of various membrane proteins in these mutant strains has been shown to be superior over that of the parental strain BL21(DE3) (Masi et al. 2003, Sorensen et al. 2003, Voet-van-Vormizeele and Groth 2003). Initial characterization of the C41(DE3) and C43(DE3) strain indicated that membrane protein production was accompanied by the proliferation of extra intracellular membranes, thereby providing the cell with extra space for inserting membrane proteins (Arechaga et al. 2000). Later on, the group of de Gier showed that C41(DE3) and C43(DE3) carry mutations in the lacUV5 promoter, which decrease the transcription of the T7 RNA polymerase and thereby down-regulate the transcription of the target gene(s) (Wagner et al. 2008a).

Recently, Bowie and coworkers showed that expression of membrane proteins in *E. coli* can be enhanced up to 90-fold. For this, two antibiotic selection markers were fused to the C-terminus of the target membrane protein. The host cells were subjected to random mutagenesis by either using the non-natural base 2-aminopurine or the mutator gene mutD5, followed by selection for resistance to the first antibiotic marker. The resistant strains were isolated and transformed with plasmid possessing the target membrane protein tagged with the second antibiotic marker and then subjected to a second round of mutagenesis and selection on the basis of both antibiotic markers (Massey-Gendel et al. 2009). The basis for the improved membrane protein overproduction characteristics of the evolved strains has not been determined. Similarly, Poolman and colleagues have isolated *L. lactis* NZ9000 strains with a two to eightfold increase in the expression of various membrane proteins. The evolution of these strains was based on a selection method using a fusion of the target protein to both GFP and an erythromycin resistance marker (ErmC) at its C-terminus. Whereas increased resistance towards erythromycin was used to select for hosts which yielded increased expression, the (increased) GFP fluorescence was used to ensure for correct folding of the target protein (Linares et al. 2010). Characterization of the strains by whole genome sequencing showed that mutations in the *nisK* gene were responsible for the improved expression levels. This gene encodes the sensor protein of a two-component regulatory system which directs nisin A-mediated expression. The improved expression in the evolved *L. lactis* strains was not due to a higher cell yield but the protein production per cell was higher.

7 Protein Engineering

Next to optimizing the expression host, the target membrane protein can also be submitted to rounds of mutagenesis to improve its stability. The poor stability of solubilized membrane proteins can be enhanced by the introduction of disulfide bonds (Lau et al. 1999, Standfuss et al. 2007). Bowie and coworkers have employed a mutagenesis approach to increase the thermostability of diacylglycerol kinase by substituting 20 amino acids with cysteine residues. Although most of the cysteine mutants had similar or higher thermal stability compared to Cysless-diacylglycerol kinase, two mutants were found to have significantly increased stability. A double mutant in which the two single mutations were combined displayed the highest thermostability (Lau et al. 1999). To enhance the thermostabilty of GPCRs, Tate and coworkers have employed an alanine-scanning approach whereby 318 residues of the β1-adrenergic receptor were individually replaced by an alanine. By combining mutations, each yielding an increased thermostability, a six-point mutant was obtained with a spectacular increase in thermostability of 21°C. This mutant showed increased solubility in various detergents and was successfully crystallized (Serrano-Vega et al. 2008, Warne et al. 2008). Recently, Plückthun and coworkers have designed an efficient strategy to engineer GPCR's for higher levels of functional expression. A periplasmic expression with cytometric screening (PECS) methodology was employed for isolation of rat neurotensin (NTR1) with improved stability and defined ligand sensitivity. A library of NTR1 mutants generated through a directed evolution approach was expressed in *E. coli* and exposed to a fluorescent substrate. The cells expressing higher levels of functional NTR1 were subsequently isolated by FACS (Sarkar et al. 2008).

8 Homologue Screening

Rather than optimizing the expression of a given protein, one may screen a set of homologous proteins and choose the best for further functional or structural studies. In the late nineties of the last century, Rees and coworkers employed this strategy for the overexpression of channel proteins. In case of the mechanosensitive channel of large conductance (MscL), homologues from nine different prokaryotic species were analyzed for expression in *E. coli* (Chen et al. 2007) and one of these ultimately proved successful in crystallization trials. Similarly, the first ABC transporter was crystallized following cloning and expressing in *E. coli* of 28 distinct ABC transporter genes (Locher et al. 2002).

9 Stress Response Upon Membrane Protein Expression

Despite increased knowledge of membrane protein biogenesis, the response of the host cell to recombinant protein production is far from clear. In 2003, Oesterhelt and colleagues monitored the Unfolded Protein Response (UPR) in *S. cerevisiae* to overexpression of membrane proteins. The host cell contained a reporter plasmid with the *lacZ* gene under the control of a transcriptional UPR element. The expression of a membrane protein, using a strong promoter and multi-copy vector, resulted in a high UPR, indicating saturation of the ER folding apparatus. The functional expression of the respective protein could be enhanced by reducing the synthesis rate of the protein to a level where the UPR was minimal. The overexpression of an endogenous protein did not lead to an UPR, suggesting that these proteins did not encounter folding problems (Griffith et al. 2003). Thus, the UPR may be used as a measure to assess the amount of folded and unfolded protein. Despite the attractiveness of the approach, the UPR has not really been exploited to screen for heterologous expression of membrane proteins in yeast.

Bill and coworkers have quantified the *S. cerevisiae* cell response to high- and low membrane protein overproduction by using yeast mini-arrays. They observed significant differential expression of 39 genes, mostly involved in membrane protein translocation and cellular physiology. Notably, the levels of transcripts for proteins involved in ribosome biogenesis, components of protein translocation and vesicular trafficking were suppressed upon membrane protein overproduction. Next to this, the increased levels of the *SPR102* transcript, encoding the β-subunit of Signal Recognition Particle (SRP), pointed towards overloading of the translocation machinery on top of the normal 10- to 100-fold excess of ribosomes over SRP (Bonander et al. 2005). Based on these observations, 43 deletion strains were constructed and their analysis demonstrated that deletion of *GCN5*, *SPT3* and *SRB5* (all involved in transcriptional regulation) enhanced the membrane protein levels. An upregulation of *BMS1* transcript was observed in these deletion strains which resulted in improved protein expression, which suggested a direct link to protein synthesis. Bms1p is a nucleolar protein that regulates the biogenesis of the 40S subunit of the ribosome. Furthermore, tuning the *BMS1* transcript levels by using doxycycline resulted in improved expression of various membrane and soluble proteins. The increased expression paralleled an adjustment of the ratio of 60S and 40S ribosomal subunits from 1:1 to 2:1, leading to slower growth and improved metabolic efficiency under high yielding conditions (Bonander et al. 2009).

The physiological response of the *E.coli* host BL21(DE3) to membrane protein expression suggests both a general stress (including heat shock) and a specific metabolic response. In the employed experiments (Wagner et al. 2007), the cells produced cytoplasmic aggregates containing the overexpressed proteins. Various chaperones and proteases were upregulated in these cells. In addition, the accumulation of periplasmic and outer membrane protein precursors in the cytoplasmic aggregates suggested a saturation of the cell translocation machinery. The levels of cytoplasmic respiratory chain complex proteins were decreased resulting in the induction of the acetate pathway for ATP production and down-regulation of the tricarboxylic acid (TCA) cycle (Wagner et al. 2007).

The characterization of the response of *L. lactis* to recombinant membrane production points towards a general and a specific stress response. Several chaperones and proteases are upregulated, presumably due to the accumulation of misfolded protein in the cell. A severe effect on cell growth was observed which is probably due to a diversion of nutrients towards the synthesis of recombinant protein or problems with the expression of endogenous membrane proteins and consequently a decreased internalization of nutrients. Unlike to the situation in *E. coli* an overloading of the protein translocation was not observed (Marreddy et al. 2010). In addition, to a more general stress and metabolic response, a specific CesSR-mediated cell envelope response was observed. In fact, a wide variety of genes involved in cell growth and membrane protein biogenesis were upregulated upon membrane protein production (Marreddy et al. 2010, Pinto et al. 2011). The forward engineering of the strains by co-expressing CesSR-regulated proteins could restore the growth as well as the expression levels of some target proteins (Pinto et al. 2011).

To understand the difference in expression levels for various proteins in *L. lactis* cells grown in complex broth (GM17) and synthetic medium (GCDM), a comprehensive and quantitative proteomic analysis was undertaken. This study indicated that several proteins involved in peptide and amino acid metabolism were upregulated in the synthetic medium under conditions that membrane protein expression was generally low, suggesting a limitation in the capacity to accumulate branched-chain amino acids in the cytoplasm. Supplementation of the synthetic medium with branched-chain amino acid-containing di-peptides or an increased expression level of the branched-chain amino acid transporter relieved the stress response and increased the protein overexpression levels (Marreddy et al. 2010).

10 Rational Engineering of Membrane Protein Expression Through Co-expression of Limiting Factors

The recombinant membrane protein production involves a coordinated action of various proteins involved in their biogenesis, i.e., translation, membrane targeting and insertion and correct folding. Often overexpressed membrane proteins accumulate either as degraded product or in inclusion bodies, which may be a direct consequence of either a lack or shortage of specific molecular chaperones or other biogenesis factors. An attractive strategy to alleviate this bottleneck is to supplement the host cell with proteins that are limiting. Even though this method is routinely employed to enhance the expression levels for challenging soluble proteins, it has proven extremely tedious to find the proper co-expressing molecule (Tolia and Joshua-Tor 2006). Tate and coworkers optimized the expression of the serotonine transporter (SERT) in insect cells by overexpressing molecular chaperones involved in protein folding in the endoplasmic reticulum (Tate et al. 1999). Out of the four molecular chaperones tested only calnexin enhanced the functional expression significantly.

Wang and coworkers attempted to increase the functional expression of a transporter in *E.coli* by co-expressing various chaperones (Chen et al. 2003). The overexpression of the bacterial magnesium transporter (CorA) in *E. coli* resulted mostly in inclusion body formation. Co-expression of chaperones (GroEL/ES), the component of signal recognition particle (Ffh) or SecA did not improve the fraction of folded CorA protein. However, the co-expression of the chaperones DnaK/J prevented the formation of inclusion bodies and enhanced the insertion of the protein into the membrane. Similarly, Georgiou and coworkers showed that co-expression of the membrane bound cell division protease FtsH could enhance the expression yields in *E. coli* of four different classes of human GPCR. Despite the increase in the expression levels, the amount of functionally active GPCR was not enhanced upon FtsH co-expression (Link et al. 2008).

In *L. lactis,* membrane protein expression elicited a specific cell envelop stress response, which is mediated by a two-component system CesSR. Deletion of *cesSR* genes, in particular *FtsH*, *OxaA2*, *llmg_2163* and *RmaB* reduced the growth of the cells as well as the expression of target membrane proteins was lowered. *In trans* complementation of these knock-out strains, using either low or high-copy plasmids, restored the growth and the production of target membrane protein. In fact, the expression levels for eukaryotic membrane proteins (PS1☆9 and StSUT1) in *L. lactis* were enhanced 2–3 fold upon co-expression of CesSR from a high-copy plasmid (Pinto et al. 2011).

The fine tuning of the co-expression of critical factors is an essential factor in the forward engineering of cells for functional protein production. In *L. lactis*, the relatively low expression level of a subset of membrane proteins in GCDM-grown cells could be restored by the co-expression of the branched chain amino acid permease (BcaP). Here, it was critical to use a low-copy number plasmid for the co-expression of BcaP as this critical factor became a burden to the cell when a high-copy plasmid was used (Marreddy et al. 2010).

11 Strategies to Optimize Functional Expression

(1) *Control of transcription rate.* An important consideration is the balance between transcription, translation and the further downstream steps such as membrane targeting, insertion and folding. Several studies have shown that generation of high amount of transcripts does not necessarily result in high amounts of functional protein (Bonander et al. 2005, Geertsma et al. 2008, Lenoir et al. 2002, Tate et al. 2003). De Gier and coworkers have shown that in the C41(DE3) and C43(DE3) strains, mutations in the lacUV5 promoter that controls the transcription of the gene coding for the T7 RNA polymerase are key to improve membrane protein overexpression, suggesting that too much mRNA and presumably too fast a synthesis of protein causes a jam at the Sec translocon (Wagner et al. 2008a). Also, it is frequently observed that lowering of the cultivation temperature can result in significant increases in the levels of functional protein (Grisshammer et al. 1993, Lewinson et al. 2008). Similar, for *P. pastoris* a decrease in the culture temperature from 30°C to 18°C increased the levels of membrane-inserted protein, while decreasing the overall expression level (Lenoir et al. 2002). Low temperature is not always favorable. The expression levels of the glycerol facilitator protein Fps1p in *S. cerevisiae* was better at higher temperatures (Bonander et al. 2005).

(2) *Nutrient availability.* The expression of membrane proteins can be not only influenced by the burden on cells protein biogenesis components but also by the capacity to accumulate nutrient. In particular for organisms like *L. lactis* (multiple amino acid and vitamin auxotroph), but also higher eukaryotes that require a rich medium for optimal growth, down-regulation of essential transport systems may slow down growth and compromise recombinant protein expression. Moreover, a rich medium typically leads to higher cell density and more protein expressed. A severe, 5–10-fold decrease, in the levels of expression of several ABC transporters was observed in *L. lactis* when cells were transferred from complex GM17 to synthetic GCDM media (Marreddy et al.

2011); the latter was required for the incorporation of amino acid analogues such as selenomethionine into proteins. Similarly, the expression levels of rabbit muscle Ca^{2+} ATPase in *S. cerevisiae* were significantly increased in rich medium (Lenoir et al. 2002). Prolonged post-induction cell growth can result in proteolytic degradation, suggesting the growth phase at which the cells needs to be harvested is also critical (Bonander et al. 2005, Marreddy et al. 2011). Occasionally, the membrane protein expression levels can be enhanced by addition of specific ligands or chemicals, like DMSO that have been claimed to assist protein folding and stabilization (Andre et al. 2006, Chloupkova et al. 2007, Weiss et al. 1998).

(3) *Gene optimization*. Large differences in codon usage of the target gene and the expression host, and thus suboptimal translation, can result in slowing of protein synthesis, premature termination of the polypeptide chain or mis-incorporation of amino acids (Kurland and Gallant 1996, Sorensen et al. 2003). The codon usage of any organism is reflected by the codon bias of the genome and the levels of cognate amino-acylated tRNAs available in the cell. In general, highly expressing proteins contain relatively more frequently used codons and vice versa. Expression of genes with rare codons in the reading frame can result in pausing of the translation as a result of ribosomal stalling. Three strategies can be utilized to overcome the problem of suboptimal codon usage: (i) random mutagenesis of the target gene sequence; (ii) chemical synthesis of a gene with optimized host cell codon usage; and (iii) co-expression of the rare tRNA genes as was done in the *E. coli* Bl21(DE3)RIPL strains from Stratagene and the Rosetta strains from Novagen. Although on first sight one would tend to optimize the codon usage by introducing the most frequently used codons (and tRNAs), some adverse effects may pop up. Ignatova and colleagues have pointed out that for production and correct folding of multidomain proteins the abundance of tRNAs in relation to codon usage is more important than codon usuage per se, and that it can be advantageous to have clusters of slow translating codons to allow the preceding domains to fold before the entire polypeptide is synthesized (Zhang et al. 2009, Zhang and Ignatova 2009). It thus seems obvious that recombinant expression may fail when slow and fast translating codons are not optimally positioned.

(4) *Fusion tags.* Frequently, the expression of membrane proteins can be optimized by fusing tags either to their N- or C-terminus. Soluble and well-expressing proteins such as maltose-binding protein (MBP), glutathione-S-transferase (GST), thioredoxin (Trx), NusA, green fluorescent protein (GFP) or peptides like Step-tag, poly His-tag are commonly used. Most of these fusion partners are commercially available and they can be also used for detecting and quantifying the protein levels and assists their affinity purification. Studies in *E. coli* have shown that MBP fusions can increase the stability of the proteins when fused to their C-terminus (Kapust and Waugh 1999). The fusion of the mystic protein from *B. subtilis* has been proposed to facilitate the insertion into the membrane of eukaryotic membrane proteins. (Roosild et al. 2005). Despite its initial promise, there is general consensus in the field that the "mystic" observations

are based on an artifact. Kunji and coworkers have shown that the overexpression of eukaryotic mitochondrial transporters in *L. lactis* can be enhanced by the addition of host cell signal peptides to the N-terminus of the target proteins (Monne et al. 2005). A similar strategy was adopted for the expression of rat neurotensin receptor in *E. coli* (Grisshammer et al. 1993) and mouse serotonin receptor in yeast cells (Weiss et al. 1995).

12 Conclusions

Over the past two decades, a steady progress has been made in the structural genomics of membrane proteins, which is reflected in the development of a wide variety of protein expression and purification tools and a significant increase in the number of crystal structures (http://blanco.biomol.uci.edu/Membrane_Proteins_xtal.html). Despite these advancements, obtaining functional and structural information of membrane proteins is still difficult with high-level functional expression forming a major hurdle. Exploration of expression space has shown that successful production of a given protein is more likely to be successful in a homologous or closed to homologous expression hosts than in a distantly-related organism. Heterologous expression can often be optimized by careful tuning of the transcript levels, gene optimization and by forward engineering of the host cells to provide more optimal targeting, membrane insertion and/or folding conditions. Major breakthroughs in membrane protein expression have been the strategy of homologue screening, and the development of selection methods to isolate evolved host cells (Miroux and Walker 1996). In addition, the development of suitable screening methods assist in the rapid monitoring of functional protein yields in a high-throughput manner (Drew et al. 2001, Geertsma et al. 2008); the GFP fusion technology has been instrumental in many aspects

Understanding the protein biogenesis machinery and the physiological response of host cells to membrane protein production is crucial for the identification of expression bottlenecks as well as to design strategies for improving the protein production yields. Recent "-omics" studies in *E. coli, L. lactis* and *S. cerevisiae* have enabled the identification of expression bottlenecks and provided clues for the forward engineering of the cells (Bonander et al. 2009, Pinto et al. 2011, Wagner et al. 2008a). Interestingly, all these studies have shown that tuning the transcript levels (either up or down) is crucial for successful expression trials. In addition, much is to be gained from the optimization of the downstream steps of membrane protein biogenesis, but again fine-tuning is critical as the pathway components may otherwise heighten the expression hurdle themselves. Prokaryotic membrane proteins pose less of problem in expression trials than eukaryotic ones, and much is still to be learned from studies aimed at probing of the expression bottlenecks.

Acknowledgments This research work was supported by the Netherlands Proteomics Centre (NPC), the European Membrane Protein Consortium EDICT, and the Netherlands Science Foundation (NWO; Chemical Sciences Top Subsidy to BP; grant number 700-56-302). We thank Gea Schuurman-Wolters for assistance with the experiments presented Fig. 2.

References

Akermoun, M., Koglin, M., Zvalova-Iooss, D., Folschweiller, N., Dowell, S.J. and Gearing, K.L. (2005) Characterization of 16 human G protein-coupled receptors expressed in baculovirus-infected insect cells. *Protein Expr. Purif.* **44**: 65–74.

Altamura, N., Capitanio, N., Bonnefoy, N., Papa, S. and Dujardin, G. (1996) The *Saccharomyces cerevisiae* OXA1 gene is required for the correct assembly of cytochrome c oxidase and oligomycin-sensitive ATP synthase. *FEBS Lett.* **382**: 111–115.

Andre, N., Cherouati, N., Prual, C., Steffan, T., Zeder-Lutz, G., Magnin, T., Pattus, F., Michel, H., Wagner, R. and Reinhart, C. (2006) Enhancing functional production of G protein-coupled receptors in *Pichia pastoris* to levels required for structural studies via a single expression screen. *Protein Sci.* **15**: 1115–1126.

Arechaga, I., Miroux, B., Karrasch, S., Huijbregts, R., de Kruijff, B., Runswick, M.J. and Walker, J.E. (2000) Characterisation of new intracellular membranes in *Escherichia coli* accompanying large scale over-production of the b subunit of F(1)F(o) ATP synthase. *FEBS Lett.* **482**: 215–219.

Aslanidis, C. and de Jong, P.J. (1990) Ligation-independent cloning of PCR products (LIC-PCR). *Nucleic Acids Res.* **18**: 6069–6074.

Bannwarth, M. and Schulz, G.E. (2003) The expression of outer membrane proteins for crystallization. *Biochim. Biophys. Acta* **1610**: 37–45.

Bartus, C.L., Jaakola, V.P., Reusch, R., Valentine, H.H., Heikinheimo, P., Levay, A., Potter, L.T., Heimo, H., Goldman, A. and Turner, G.J. (2003) Downstream coding region determinants of bacterio-opsin, muscarinic acetylcholine receptor and adrenergic receptor expression in Halobacterium salinarum. *Biochim. Biophys. Acta* **1610**: 109–123.

Basile, G. and Peticca, H. (2009) Recombinant protein expression in *Leishmania tarentolae*. *Mol. Biotechnol.* **43**: 273–278.

Berger, I., Fitzgerald, D.J. and Richmond, T.J. (2004) Baculovirus expression system for heterologous multiprotein complexes. *Nat. Biotechnol.* **22**: 1583–1587.

Berntsson, R.P., Alia, O.N., Fusetti, F., Thunnissen, A.M., Poolman, B. and Slotboom, D.J. (2009) Selenomethionine incorporation in proteins expressed in *Lactococcus lactis*. *Protein Sci.* **18**: 1121–1127.

Bill, R.M. (2001) Yeast – a panacea for the structure-function analysis of membrane proteins? *Curr. Genet.* **40**: 157–171.

Bonander, N., Hedfalk, K., Larsson, C., Mostad, P., Chang, C., Gustafsson, L. and Bill, R.M. (2005) Design of improved membrane protein production experiments: quantitation of the host response. *Protein Sci.* **14**: 1729–1740.

Bonander, N., Darby, R.A., Grgic, L., Bora, N., Wen, J., Brogna, S., Poyner, D.R., O'Neill, M.A. and Bill, R.M. (2009) Altering the ribosomal subunit ratio in yeast maximizes recombinant protein yield. *Microb. Cell Fact.* **8**: 10.

Bongers, R.S., Veening, J.W., Van Wieringen, M., Kuipers, O.P. and Kleerebezem, M. (2005) Development and characterization of a subtilin-regulated expression system in Bacillus subtilis: strict control of gene expression by addition of subtilin. *Appl. Environ. Microbiol.* **71**: 8818–8824.

Boon, J.M. and Smith, B.D. (2002) Chemical control of phospholipid distribution across bilayer membranes. *Med. Res. Rev.* **22**: 251–281.

Breitling, R., Klingner, S., Callewaert, N., Pietrucha, R., Geyer, A., Ehrlich, G., Hartung, R., Muller, A., Contreras, R., Beverley, S.M. and Alexandrov, K. (2002) Non-pathogenic trypanosomatid protozoa as a platform for protein research and production. *Protein Expr. Purif.* **25**: 209–218.

Chebolu, S. and Daniell, H. (2009) Chloroplast-derived vaccine antigens and biopharmaceuticals: expression, folding, assembly and functionality. *Curr. Top. Microbiol. Immunol.* **332**: 33–54.

Chen, M., Samuelson, J.C., Jiang, F., Muller, M., Kuhn, A. and Dalbey, R.E. (2002) Direct interaction of YidC with the Sec-independent Pf3 coat protein during its membrane protein insertion. *J. Biol. Chem.* **277**: 7670–7675.

Chen, Y., Song, J., Sui, S.F. and Wang, D.N. (2003) DnaK and DnaJ facilitated the folding process and reduced inclusion body formation of magnesium transporter CorA overexpressed in *Escherichia coli*. *Protein Expr. Purif.* **32**: 221–231.

Chen, Y.J., Pornillos, O., Lieu, S., Ma, C., Chen, A.P. and Chang, G. (2007) X-ray structure of EmrE supports dual topology model. *Proc. Natl. Acad. Sci. USA* **104**: 18999–19004.

Chloupkova, M., Pickert, A., Lee, J.Y., Souza, S., Trinh, Y.T., Connelly, S.M., Dumont, M.E., Dean, M. and Urbatsch, I.L. (2007) Expression of 25 human ABC transporters in the yeast *Pichia pastoris* and characterization of the purified ABCC3 ATPase activity. *Biochemistry* **46**: 7992–8003.

Dalbey, R.E. and Kuhn, A. (2004) YidC family members are involved in the membrane insertion, lateral integration, folding, and assembly of membrane proteins. *J. Cell Biol.* **166**: 769–774.

Daniell, H., Streatfield, S.J. and Wycoff, K. (2001) Medical molecular farming: production of antibodies, biopharmaceuticals and edible vaccines in plants. *Trends Plant Sci.* **6**: 219–226.

Daniell, H., Khan, M.S. and Allison, L. (2002) Milestones in chloroplast genetic engineering: an environmentally friendly era in biotechnology. *Trends Plant Sci.* **7**: 84–91.

Davies, A.H. (1994) Current methods for manipulating baculoviruses. *Biotechnology (NY)* **12**: 47–50.

de Groot, B.L., Engel, A. and Grubmuller, H. (2001) A refined structure of human aquaporin-1. *FEBS Lett.* **504**: 206–211.

de Marco. A. (2007) Protocol for preparing proteins with improved solubility by co-expressing with molecular chaperones in *Escherichia coli*. *Nat. Protoc.* **2**: 2632–2639.

de Ruyter, P.G., Kuipers, O.P. and de Vos, W.M. (1996) Controlled gene expression systems for *Lactococcus lactis* with the food-grade inducer nisin. *Appl. Environ. Microbiol.* **62**: 3662–3667.

de Smet. L., Kostanjevecki, V., Guisez, Y. and Van Beeumen, J. (2001) A novel system for heterologous expression of flavocytochrome c in phototrophic bacteria using the *Allochromatium vinosum* rbcA promoter. *Arch. Microbiol.* **176**: 19–28.

de Vos, W.M. (1999) Gene expression systems for lactic acid bacteria. *Curr. Opin. Microbiol.* **2**:289–295.

Drew, D., Sjostrand, D., Nilsson, J., Urbig, T., Chin, C.N., de Gier, J.W. and von, H.H. (2002) Rapid topology mapping of *Escherichia coli* inner-membrane proteins by prediction and PhoA/GFP fusion analysis. *Proc. Natl. Acad. Sci. USA* **99**: 2690–2695.

Drew, D., Lerch, M., Kunji, E., Slotboom, D.J. and de Gier, J.W. (2006) Optimization of membrane protein overexpression and purification using GFP fusions. *Nat. Methods* **3**: 303–313.

Drew, D.E., G. von, H., Nordlund, P. and de Gier, J.W. (2001) Green fluorescent protein as an indicator to monitor membrane protein overexpression in *Escherichia coli*. *FEBS Lett.* **507**: 220–224.

Drew, D., Newstead, S., Sonoda, Y., Kim, H., von, H.G. and Iwata, S. (2008) GFP-based optimization scheme for the overexpression and purification of eukaryotic membrane proteins in *Saccharomyces cerevisiae*. *Nat. Protoc.* **3**: 784–798.

Du, D., Kato, T., Suzuki, F. and Park, E.Y. (2009) Expression of protein complex comprising the human prorenin and (pro)renin receptor in silkworm larvae using *Bombyx mori* nucleopolyhedrovirus (BmNPV) bacmids for improving biological function. *Mol. Biotechnol.* **43**: 154–161.

El, K.M., van Roosmalen, M.L., Jager, D., Metselaar, H., Permentier, H., Leenhouts, K. and Broos, J. (2008) *Lactococcus lactis* as expression host for the biosynthetic incorporation of tryptophan analogues into recombinant proteins. *Biochem. J.* **409**: 193–198.

Elias, C.B., Jardin, B. and Kamen, A. (2007) Recombinant protein production in large-scale agitated bioreactors using the baculovirus expression vector system. *Methods Mol. Biol.* **388**: 225–246.

Felder, C.C., Veluz, J.S., Williams, H.L., Briley, E.M. and Matsuda, L.A. (1992) Cannabinoid agonists stimulate both receptor- and non-receptor-mediated signal transduction pathways in cells transfected with and expressing cannabinoid receptor clones. *Mol. Pharmacol.* **42**: 838–845.

Frelet-Barrand, A., Boutigny, S., Moyet, L., Deniaud, A., Seigneurin-Berny, D., Salvi, D., Bernaudat, F., Richaud, P., Pebay-Peyroula, E., Joyard, J. and Rolland, N. (2010) *Lactococcus lactis*, an alternative system for functional expression of peripheral and intrinsic Arabidopsis membrane proteins. *PLoS. One.* **5**: e8746.

Funes, S., Hasona, A., Bauerschmitt, H., Grubbauer, C., Kauff, K., Collins, R., Crowley, P.J., Palmer, S.R., Brady, L.J. and Herrmann, J.M. (2009) Independent gene duplications of the YidC/Oxa/Alb3 family enabled a specialized cotranslational function. *Proc. Natl. Acad. Sci. USA* **106**: 6656–6661.

Geertsma, E.R. and Poolman, B. (2007) High-throughput cloning and expression in recalcitrant bacteria. *Nat. Methods* **4**: 705–707.

Geertsma, E.R., Groeneveld, M., Slotboom, D.J. and Poolman, B. (2008) Quality control of overexpressed membrane proteins. *Proc. Natl. Acad. Sci. USA* **105**: 5722–5727.

Geissendorfer, M. and Hillen, W. (1990) Regulated expression of heterologous genes in Bacillus subtilis using the Tn10 encoded tet regulatory elements. *Appl. Microbiol. Biotechnol.* **33**: 657–663.

Gomez, N., Carrillo, C., Salinas, J., Parra, F., Borca, M.V. and Escribano, J.M. (1998) Expression of immunogenic glycoprotein S polypeptides from transmissible gastroenteritis coronavirus in transgenic plants. *Virology* **249**: 352–358.

Gomord, V. and Faye, L. (2004) Posttranslational modification of therapeutic proteins in plants. *Curr. Opin. Plant Biol.* **7**: 171–181.

Gonzales, E.B., Kawate, T. and Gouaux, E. (2009) Pore architecture and ion sites in acid-sensing ion channels and P2X receptors. *Nature* **460**: 599–604.

Gonzalez-Montalban, N., Garcia-Fruitos, E. and Villaverde, A. (2007) Recombinant protein solubility – does more mean better? *Nat. Biotechnol.* **25**: 718–720.

Gordon, E., Horsefield, R., Swarts, H.G., de Pont, J.J., Neutze, R. and Snijder, A. (2008) Effective high-throughput overproduction of membrane proteins in *Escherichia coli*. *Protein Expr. Purif.* **62**: 1–8.

Griffith, D.A., Delipala, C., Leadsham, J., Jarvis, S.M. and Oesterhelt, D. (2003) A novel yeast expression system for the overproduction of quality-controlled membrane proteins. *FEBS Lett.* **553**: 45–50.

Grisshammer, R., Duckworth, R. and Henderson, R. (1993) Expression of a rat neurotensin receptor in *Escherichia coli*. *Biochem. J.* **295**(Pt 2): 571–576.

Grisshammer, R. and Tate, C.G. (1995) Overexpression of integral membrane proteins for structural studies. *Q. Rev. Biophys.* **28**: 315–422.

Grisshammer, R. (2006) Understanding recombinant expression of membrane proteins. *Curr. Opin. Biotechnol.* **17**: 337–340.

Hannig, G. and Makrides, S.C. (1998) Strategies for optimizing heterologous protein expression in *Escherichia coli*. *Trends Biotechnol.* **16**: 54–60.

Haq, T.A., Mason, H.S., Clements, J.D. and Arntzen, C.J. (1995) Oral immunization with a recombinant bacterial antigen produced in transgenic plants. *Science* **268**: 714–716.

Harwood, S. (2007) Small-scale protein production with the baculovirus expression vector system. *Methods Mol. Biol.* **388**: 211–224.

Hasona, A., Crowley, P.J., Levesque, C.M., Mair, R.W., Cvitkovitch, D.G., Bleiweis, A.S. and Brady, L.J. (2005) Streptococcal viability and diminished stress tolerance in mutants lacking the signal recognition particle pathway or YidC2. *Proc. Natl. Acad. Sci. USA* **102**: 17466–17471.

Helenius, A. and Aebi, M. (2004) Roles of N-linked glycans in the endoplasmic reticulum. *Annu. Rev. Biochem.* **73**: 1019–1049.

Hiroaki, Y., Tani, K., Kamegawa, A., Gyobu, N., Nishikawa, K., Suzuki, H., Walz, T., Sasaki, S., Mitsuoka, K., Kimura, K., Mizoguchi, A. and Fujiyoshi, Y. (2006) Implications of the aquaporin-4 structure on array formation and cell adhesion. *J. Mol. Biol.* **355**: 628–639.

Hollenberg, C.P. and Gellissen, G. (1997) Production of recombinant proteins by methylotrophic yeasts. *Curr. Opin. Biotechnol.* **8**: 554–560.

Huber, D., Boyd, D., Xia, Y., Olma, M.H., Gerstein, M. and Beckwith, J. (2005) Use of thioredoxin as a reporter to identify a subset of *Escherichia coli* signal sequences that promote signal recognition particle-dependent translocation. *J. Bacteriol.* **187**: 2983–2991.

Jacobson, A. and Peltz, S.W. (1996) Interrelationships of the pathways of mRNA decay and translation in eukaryotic cells. *Annu. Rev. Biochem.* **65**: 693–739.

Jarvie, K.R., Tiberi, M., Silvia, C., Gingrich, J.A. and Caron, M.G. (1993) Molecular cloning, stable expression and desensitization of the human dopamine D1b/D5 receptor. *J. Recept. Res.* **13**: 573–590.

Jidenko, M., Nielsen, R.C., Sorensen, T.L., Moller, J.V., Le, M.M., Nissen, P. and Jaxel, C. (2005) Crystallization of a mammalian membrane protein overexpressed in *Saccharomyces cerevisiae*. *Proc. Natl. Acad. Sci. USA* **102**: 11687–11691.

Jin, R., Singh, S.K., Gu, S., Furukawa, H., Sobolevsky, A.I., Zhou, J., Jin, Y. and Gouaux, E. (2009) Crystal structure and association behaviour of the GluR2 amino-terminal domain. *EMBO J.* **28**: 1812–1823.

Jordan, P., Fromme, P., Witt, H.T., Klukas, O., Saenger, W. and Krauss, N. (2001) Three-dimensional structure of cyanobacterial photosystem I at 2.5 A resolution. *Nature* **411**: 909–917.

Kalmbach, R., Chizhov, I., Schumacher, M.C., Friedrich, T., Bamberg, E. and Engelhard, M. (2007) Functional cell-free synthesis of a seven helix membrane protein: in situ insertion of bacteriorhodopsin into liposomes. *J. Mol. Biol.* **371**: 639–648.

Kappler, U. and McEwan, A.G. (2002) A system for the heterologous expression of complex redox proteins in Rhodobacter capsulatus: characterisation of recombinant sulphite:cytochrome c oxidoreductase from Starkeya novella. *FEBS Lett.* **529**: 208–214.

Kapust, R.B. and Waugh, D.S. (1999) *Escherichia coli* maltose-binding protein is uncommonly effective at promoting the solubility of polypeptides to which it is fused. *Protein Sci.* **8**: 1668–1674.

Kawate, T., Michel, J.C., Birdsong, W.T and Gouaux, E. (2009) Crystal structure of the ATP-gated P2X(4) ion channel in the closed state. *Nature* **460**: 592–598.

Keynan, S., Suh, Y.J., Kanner, B.I. and Rudnick, G. (1992) Expression of a cloned gamma-aminobutyric acid transporter in mammalian cells. *Biochemistry* **31**: 1974–1979.

Kiefer, H. (2003) In vitro folding of alpha-helical membrane proteins. *Biochim. Biophys. Acta* **1610**: 57–62.

Kigawa, T., Yabuki, T., Matsuda, N., Matsuda, T., Nakajima, R., Tanaka, A. and Yokoyama, T. (2004) Preparation of *Escherichia coli* cell extract for highly productive cell-free protein expression. *J. Struct. Funct. Genomics* **5**: 63–68.

Kim, D.M. and Choi, C.Y. (1996) A semicontinuous prokaryotic coupled transcription//translation system using a dialysis membrane. *Biotechnol. Prog.* **12**: 645–649.

Klaassen, C.H. and DeGrip, W.J. (2000) Baculovirus expression system for expression and characterization of functional recombinant visual pigments. *Methods Enzymol.* **315**: 12–29.

Koch, H.G., Hengelage, T., Neumann-Haefelin, C., MacFarlane, J., Hoffschulte, H.K., Schimz, K.L., Mechler, B. and Muller, M. (1999) In vitro studies with purified components reveal signal recognition particle (SRP) and SecA/SecB as constituents of two independent protein-targeting pathways of *Escherichia coli. Mol. Biol. Cell* **10**: 2163–2173.

Korepanova, A., Gao, F.P., Hua, Y., Qin, H., Nakamoto, R.K. and Cross, T.A. (2005) Cloning and expression of multiple integral membrane proteins from *Mycobacterium tuberculosis* in *Escherichia coli*. *Protein Sci.* **14**: 148–158.

Kost, T.A., Condreay, J.P. and Jarvis, D.L. (2005) Baculovirus as versatile vectors for protein expression in insect and mammalian cells. *Nat. Biotechnol.* **23**: 567–575.

Kunji, E.R., Slotboom, D.J. and Poolman, B. (2003) *Lactococcus lactis* as host for overproduction of functional membrane proteins. *Biochim. Biophys. Acta* **1610**: 97–108.

Kurland, C. and Gallant, J. (1996) Errors of heterologous protein expression. *Curr. Opin. Biotechnol.* **7**: 489–493.

Kushnir, S., Gase, K., Breitling, R. and Alexandrov, K. (2005) Development of an inducible protein expression system based on the protozoan host *Leishmania tarentolae*. *Protein Expr. Purif.* **42**: 37–46.

Lau, F.W., Nauli, S., Zhou, Y. and Bowie, J.U. (1999) Changing single side-chains can greatly enhance the resistance of a membrane protein to irreversible inactivation. *J. Mol. Biol.* **290**: 559–564.

Lenoir, G., Menguy, T., Corre, F., Montigny, C., Pedersen, P.A., Thines, D., Le, M.M. and Falson, P. (2002) Overproduction in yeast and rapid and efficient purification of the rabbit SERCA1a Ca(2+)-ATPase. *Biochim. Biophys. Acta* **1560**: 67–83.

Lewinson, O., Lee, A.T. and Rees, D.C. (2008) The funnel approach to the precrystallization production of membrane proteins. *J. Mol. Biol.* **377**: 62–73.

Li, M., Hays, F.A., Roe-Zurz, Z., Vuong, L., Kelly, L., Ho, C.M., Robbins, R.M., Pieper, U., O'Connell, J.D., III, Miercke, L.J., Giacomini, K.M., Sali, A. and Stroud, R.M. (2009) Selecting optimum eukaryotic integral membrane proteins for structure determination by rapid expression and solubilization screening. *J. Mol. Biol.* **385**: 820–830.

Li, Y., Kijac, A.Z., Sligar, S.G. and Rienstra, C.M. (2006) Structural analysis of nanoscale self-assembled discoidal lipid bilayers by solid-state NMR spectroscopy. *Biophys. J.* **91**: 3819–3828.

Liang, B. and Tamm, L.K. (2007) Structure of outer membrane protein G by solution NMR spectroscopy. *Proc. Natl. Acad. Sci. USA* **104**: 16140–16145.

Linares, D.M., Geertsma, E.R. and Poolman, B. (2010) Evolved *Lactococcus lactis* strains for enhanced expression of recombinant membrane proteins. *J. Mol. Biol.* **401**: 45–55.

Link, A.J., Skretas, G., Strauch, E.M., Chari, N.S. and Georgiou, G. (2008) Efficient production of membrane-integrated and detergent-soluble G protein-coupled receptors in *Escherichia coli*. *Protein Sci.* **17**: 1857–1863.

Liu, Q., Li, M.Z., Leibham, D., Cortez, D. and Elledge, S.J. (1998) The univector plasmid-fusion system, a method for rapid construction of recombinant DNA without restriction enzymes. *Curr. Biol.* **8**: 1300–1309.

Llull, D. and Poquet, I. (2004) New expression system tightly controlled by zinc availability in *Lactococcus lactis*. *Appl. Environ. Microbiol.* **70**: 5398–5406.

Locher, K.P., Lee, A.T. and Rees, D.C. (2002) The *E. coli* BtuCD structure: a framework for ABC transporter architecture and mechanism. *Science* **296**: 1091–1098.

Luckow, V.A. (1993) Baculovirus systems for the expression of human gene products. *Curr. Opin. Biotechnol.* **4**: 564–572.

Luckow, V.A., Lee, S.C., Barry, G.F. and Olins, P.O. (1993) Efficient generation of infectious recombinant baculoviruses by site-specific transposon-mediated insertion of foreign genes into a baculovirus genome propagated in *Escherichia coli*. *J. Virol.* **67**: 4566–4579.

Luecke, H., Richter, H.T. and Lanyi, J.K. (1998) Proton transfer pathways in bacteriorhodopsin at 2.3 angstrom resolution. *Science* **280**: 1934–1937.

Madin, K., Sawasaki, T., Ogasawara, T. and Endo, Y. (2000) A highly efficient and robust cell-free protein synthesis system prepared from wheat embryos: plants apparently contain a suicide system directed at ribosomes. *Proc. Natl. Acad. Sci. USA* **97**: 559–564.

Maliga, P. (2002) Engineering the plastid genome of higher plants. *Curr. Opin. Plant Biol.* **5**: 164–172.

Mancia, F., Patel, S.D., Rajala, R.W., Scherer, P.E., Nemes, A., Schieren, I., Hendrickson, W.A. and Shapiro, L. (2004) Optimization of protein production in mammalian cells with a coexpressed fluorescent marker. *Structure.* **12**: 1355–1360.

Mancia, F. and Hendrickson, W.A. (2007) Expression of recombinant G-protein coupled receptors for structural biology. *Mol. Biosyst.* **3**: 723–734.

Marreddy, R.K., Geertsma, E.R., Permentier, H.P., Pinto, J.P., Kok, J. and Poolman, B. (2010) Amino acid accumulation limits the overexpression of proteins in *Lactococcus lactis*. *PLoS ONE*. **5**: e10317.

Marreddy, R.K., Pinto, J.P., Wolters, J.C., Geertsma, E.R., Fusetti, F., Permentier, H., Kuipers, O.P., Kok, J. and Poolman, B. (2011) The response of *Lactococcus lactis* to membrane protein production. Under evaluation.

Masi, M., Pages, J.M. and Pradel, E. (2003) Overexpression and purification of the three components of the *Enterobacter aerogenes* AcrA-AcrB-TolC multidrug efflux pump. *J. Chromatogr. B Analyt. Technol. Biomed. Life Sci.* **786**: 197–205.

Mason, H.S., Lam, D.M. and Arntzen, C.J. (1992) Expression of hepatitis B surface antigen in transgenic plants. *Proc. Natl. Acad. Sci. USA* **89**: 11745–11749.

Massey-Gendel, E., Zhao, A., Boulting, G., Kim, H.Y., Balamotis, M.A., Seligman, L.M., Nakamoto, R.K. and Bowie, J.U. (2009) Genetic selection system for improving recombinant membrane protein expression in *E. coli*. *Protein Sci.* **18**: 372–383.

Miroux, B. and Walker, J.E. (1996) Over-production of proteins in *Escherichia coli*: mutant hosts that allow synthesis of some membrane proteins and globular proteins at high levels. *J. Mol. Biol.* **260**: 289–298.

Miyoshi, A., Jamet, E., Commissaire, J., Renault, P., Langella, P. and Azevedo, V. (2004) A xylose-inducible expression system for *Lactococcus lactis*. *FEMS Microbiol. Lett.* **239**: 205–212.

Monne, M., Chan, K.W., Slotboom, D.J. and Kunji, E.R. (2005) Functional expression of eukaryotic membrane proteins in *Lactococcus lactis*. *Protein Sci.* **14**: 3048–3056.

Moore, M., Harrison, M.S., Peterson, E.C. and Henry, R. (2000) Chloroplast Oxa1p homolog albino3 is required for post-translational integration of the light harvesting chlorophyll-binding protein into thylakoid membranes. *J. Biol. Chem.* **275**: 1529–1532.

Morth, J.P., Pedersen, B.P., Toustrup-Jensen, M.S., Sorensen, T.L., Petersen, J., Andersen, J.P., Vilsen, B. and Nissen, P. (2007) Crystal structure of the sodium-potassium pump. *Nature* **450**: 1043–1049.

Mulligan, C., Geertsma, E.R., Severi, E., Kelly, D.J., Poolman, B. and Thomas, G.H. (2009) The substrate-binding protein imposes directionality on an electrochemical sodium gradient-driven TRAP transporter. *Proc. Natl. Acad. Sci. USA* **106**: 1778–1783.

Murata, K., Mitsuoka, K., Hirai, T., Walz, T., Agre, P., Heymann, J.B., Engel, A. and Fujiyoshi, Y. (2000) Structural determinants of water permeation through aquaporin-1. *Nature* **407**:599–605.

Nagamori, S., Smirnova, I.N. and Kaback, H.R. (2004) Role of YidC in folding of polytopic membrane proteins. *J. Cell Biol.* **165**: 53–62.

Newbury, S.F., Smith, N.H., Robinson, E.C., Hiles, I.D. and Higgins, C.F. (1987) Stabilization of translationally active mRNA by prokaryotic REP sequences. *Cell* **48**: 297–310.

Newstead, S., Kim, H., von, H.G., Iwata, S. and Drew, D. (2007) High-throughput fluorescent-based optimization of eukaryotic membrane protein overexpression and purification in *Saccharomyces cerevisiae*. *Proc. Natl. Acad. Sci. USA* **104**: 13936–13941.

Ng, D.T., Brown, J.D. and Walter, P. (1996) Signal sequences specify the targeting route to the endoplasmic reticulum membrane. *J. Cell Biol.* **134**: 269–278.

Niebauer, R.T., Wedekind, A. and Robinson, A.S. (2004) Decreases in yeast expression yields of the human adenosine A2a receptor are a result of translational or post-translational events. *Protein Expr. Purif.* **37**: 134–143.

Niu, Y., Kong, J. and Xu, Y. (2008) A novel GFP-fused eukaryotic membrane protein expression system in *Lactococcus lactis* and its application to overexpression of an Elongase. *Curr. Microbiol.* **57**: 423–428.

Nomura, S.M., Kondoh, S., Asayama, W., Asada, A., Nishikawa, S. and Akiyoshi, K. (2008) Direct preparation of giant proteo-liposomes by in vitro membrane protein synthesis. *J. Biotechnol.* **133**: 190–195.

Nozawa, A., Nanamiya, H., Miyata, T., Linka, N., Endo, Y., Weber, A.O. and Tozawa, Y. (2007) A cell-free translation and proteoliposome reconstitution system for functional analysis of plant solute transporters. *Plant Cell Physiol* **48**: 1815–1820.

Olesen, C., Picard, M., Winther, A.M., Gyrup, C., Morth, J.P., Oxvig, C., Moller, J.V. and Nissen, P. (2007) The structural basis of calcium transport by the calcium pump. *Nature* **450**: 1036–1042.

Opekarova, M. and Tanner, W. (2003) Specific lipid requirements of membrane proteins – a putative bottleneck in heterologous expression. *Biochim. Biophys. Acta* **1610**: 11–22.

Osterberg, M., Kim, H., Warringer, J., Melen, K., Blomberg, A. and von, H.G (2006) Phenotypic effects of membrane protein overexpression in *Saccharomyces cerevisiae*. *Proc. Natl. Acad. Sci. USA* **103**: 11148–11153.

Palanivelu, D.V., Kozono, D.E., Engel, A., Suda, K., Lustig, A., Agre, P. and Schirmer, T. (2006) Co-axial association of recombinant eye lens aquaporin-0 observed in loosely packed 3D crystals. *J. Mol. Biol.* **355**: 605–611.

Palczewski, K., Kumasaka, T., Hori, T., Behnke, C.A., Motoshima, H., Fox, B.A., Le, I, T., Teller, D.C., Okada, T., Stenkamp, R.E., Yamamoto, M. and Miyano, M. (2000) Crystal structure of rhodopsin: a G protein-coupled receptor. *Science* **289**: 739–745.

Pinto, J.P., Kuipers, O.P., Marreddy, R.K., Poolman, B. and Kok, J. (2011) Efficient overproduction of membrane proteins in *Lactococcus lactis* relies on the cell envelope stress sensor/regulator couple CesSR. Under evaluation.

Quick, M. and Wright, E.M. (2002) Employing *Escherichia coli* to functionally express, purify, and characterize a human transporter. *Proc. Natl. Acad. Sci. USA* **99**: 8597–8601.

Quick, M. and Javitch, J.A. (2007) Monitoring the function of membrane transport proteins in detergent-solubilized form. *Proc. Natl. Acad. Sci. USA* **104**: 3603–3608.

Rapoport, T.A. (2007) Protein translocation across the eukaryotic endoplasmic reticulum and bacterial plasma membranes. *Nature* **450**: 663–669.

Reuben, M., Rising, L., Prinz, C., Hersey, S. and Sachs, G. (1994) Cloning and expression of the rabbit gastric CCK-A receptor. *Biochim. Biophys. Acta* **1219**: 321–327.

Roosild, T.P., Greenwald, J., Vega, M., Castronovo, S., Riek, R. and Choe, S. (2005) NMR structure of Mistic, a membrane-integrating protein for membrane protein expression. *Science* **307**: 1317–1321.

Sarkar, C.A., Dodevski, I., Kenig, M., Dudli, S., Mohr, A., Hermans, E. and Pluckthun, A. (2008) Directed evolution of a G protein-coupled receptor for expression, stability, and binding selectivity. *Proc. Natl. Acad. Sci. USA* **105**: 14808–14813.

Savage, D.F., Anderson, C.L., Robles-Colmenares, Y., Newby, Z.E. and Stroud, R.M. (2007) Cell-free complements in vivo expression of the *E. coli* membrane proteome. *Protein Sci.* **16**: 966–976.

Sawasaki, T., Hasegawa, Y., Tsuchimochi, M., Kamura, N., Ogasawara, T., Kuroita, T. and Endo, Y. (2002) A bilayer cell-free protein synthesis system for high-throughput screening of gene products. *FEBS Lett.* **514**: 102–105.

Sawasaki, T., Gouda, M.D., Kawasaki, T., Tsuboi, T., Tozawa, Y., Takai, K. and Endo, Y. (2005) The wheat germ cell-free expression system: methods for high-throughput materialization of genetic information. *Methods Mol. Biol.* **310**: 131–144.

Schahs, M., Strasser, R., Stadlmann, J., Kunert, R., Rademacher, T. and Steinkellner, H. (2007) Production of a monoclonal antibody in plants with a humanized N-glycosylation pattern. *Plant Biotechnol. J.* **5**: 657–663.

Schumann, W. (2007) Production of recombinant proteins in Bacillus subtilis. *Adv. Appl. Microbiol.* **62**: 137–189.

Schunemann, D. (2004) Structure and function of the chloroplast signal recognition particle. *Curr. Genet.* **44**: 295–304.

Schurmann, A., Monden, I., Joost, H.G. and Keller, K. (1992) Subcellular distribution and activity of glucose transporter isoforms GLUT1 and GLUT4 transiently expressed in COS-7 cells. *Biochim. Biophys. Acta* **1131**:245–252.

Schwarz, D., Junge, F., Durst, F., Frolich, N., Schneider, B., Reckel, S., Sobhanifar, S., Dotsch, V. and Bernhard, F. (2007) Preparative scale expression of membrane proteins in *Escherichia coli*-based continuous exchange cell-free systems. *Nat. Protoc.* **2**: 2945–2957.

Scotti, P.A., Urbanus, M.L., Brunner, J., de Gier, J.W., von, H.G., van der Does, C., Driessen, A.J., Oudega, B. and Luirink, J. (2000) YidC, the *Escherichia coli* homologue of mitochondrial Oxa1p, is a component of the Sec translocase. *EMBO J.* **19**: 542–549.

Serek, J., Bauer-Manz, G., Struhalla, G., van den, B.L., Kiefer, D., Dalbey, R. and Kuhn, A. (2004) *Escherichia coli* YidC is a membrane insertase for Sec-independent proteins. *EMBO J.* **23**: 294–301.

Serrano-Vega, M.J., Magnani, F., Shibata, Y. and Tate, C.G. (2008) Conformational thermostabilization of the beta1-adrenergic receptor in a detergent-resistant form. *Proc. Natl. Acad. Sci. USA* **105**: 877–882.

Shukla, A.K., Haase, W., Reinhart, C. and Michel, H. (2007) Heterologous expression and characterization of the recombinant bradykinin B2 receptor using the methylotrophic yeast *Pichia pastoris*. *Protein Expr. Purif.* **55**: 1–8.

Sobolevsky, A.I., Rosconi, M.P. and Gouaux, E. (2009) X-ray structure, symmetry and mechanism of an AMPA-subtype glutamate receptor. *Nature* **462**: 745–756.

Sokolova, O., Kolmakova-Partensky, L. and Grigorieff, N. (2001) Three-dimensional structure of a voltage-gated potassium channel at 2.5 nm resolution. *Structure* **9**: 215–220.

Sorensen, H.P., Sperling-Petersen, H.U. and Mortensen, K.K. (2003) Production of recombinant thermostable proteins expressed in *Escherichia coli*: completion of protein synthesis is the bottleneck. *J. Chromatogr. B Analyt. Technol. Biomed. Life Sci.* **786**: 207–214.

Sreekrishna, K., Brankamp, R.G., Kropp, K.E., Blankenship, D.T., Tsay, J.T., Smith, P.L., Wierschke, J.D., Subramaniam, A. and Birkenberger, L.A. (1997) Strategies for optimal synthesis and secretion of heterologous proteins in the methylotrophic yeast. *Pichia Pastoris. Gene* **190**: 55–62.

Standfuss, J., Xie, G., Edwards, P.C., Burghammer, M., Oprian, D.D. and Schertler, G.F. (2007) Crystal structure of a thermally stable rhodopsin mutant. *J. Mol. Biol.* **372**: 1179–1188.

Stock, D., Leslie, A.G. and Walker, J.E. (1999) Molecular architecture of the rotary motor in ATP synthase. *Science* **286**: 1700–1705.

Streatfield, S.J. (2007) Approaches to achieve high-level heterologous protein production in plants. *Plant Biotechnol. J.* **5**: 2–15.

Summers, M.D. (2006) Milestones leading to the genetic engineering of baculoviruses as expression vector systems and viral pesticides. *Adv. Virus Res.* **68**: 3–73.

Surade, S., Klein, M., Stolt-Bergner, P.C., Muenke, C., Roy, A. and Michel, H. (2006) Comparative analysis and "expression space" coverage of the production of prokaryotic membrane proteins for structural genomics. *Protein Sci.* **15**: 2178–2189.

Tackaberry, E.S., Dudani, A.K., Prior, F., Tocchi, M., Sardana, R., Altosaar, I. and Ganz, P.R. (1999) Development of biopharmaceuticals in plant expression systems: cloning, expression and immunological reactivity of human cytomegalovirus glycoprotein B (UL55) in seeds of transgenic tobacco. *Vaccine* **17**: 3020–3029.

Takegawa, K., Tohda, H., Sasaki, M., Idiris, A., Ohashi, T., Mukaiyama, H., Giga-Hama, Y. and Kumagai, H. (2009) Production of heterologous proteins using the fission-yeast (*Schizosaccharomyces pombe*) expression system. *Biotechnol.* Appl. *Biochem.* **53**: 227–235.

Tate, C.G. and Grisshammer, R. (1996) Heterologous expression of G-protein-coupled receptors. *Trends Biotechnol.* **14**: 426–430.

Tate, C.G., Whiteley, E. and Betenbaugh, M.J. (1999) Molecular chaperones stimulate the functional expression of the cocaine-sensitive serotonin transporter. *J. Biol. Chem.* **274**: 17551–17558.

Tate, C.G. (2001) Overexpression of mammalian integral membrane proteins for structural studies. *FEBS Lett.* **504**: 94–98.

Tate, C.G., Haase, J., Baker, C., Boorsma, M., Magnani, F., Vallis, Y. and Williams, D.C. (2003) Comparison of seven different heterologous protein expression systems for the production of the serotonin transporter. *Biochim. Biophys. Acta* **1610**: 141–153.

Terpe, K. (2006) Overview of bacterial expression systems for heterologous protein production: from molecular and biochemical fundamentals to commercial systems. *Appl. Microbiol. Biotechnol.* **72**: 211–222.

Thuy Le, A.T. and Schumann, W. (2007) A novel cold-inducible expression system for Bacillus subtilis. *Protein Expr. Purif.* **53**: 264–269.

Tolia, N.H. and Joshua-Tor, L. (2006) Strategies for protein coexpression in *Escherichia coli*. *Nat. Methods* **3**: 55–64.

Torizawa, T., Shimizu, M., Taoka, M., Miyano, H. and Kainosho, M. (2004) Efficient production of isotopically labeled proteins by cell-free synthesis: a practical protocol. *J. Biomol. NMR* **30**: 311–325.

Trometer, C. and Falson, P. (2010) Mammalian membrane protein expression in baculovirus-infected insect cells. *Methods Mol. Biol.* **601**: 105–117.

Turner, G.J., Miercke, L.J., Mitra, A.K., Stroud, R.M., Betlach, M.C. and Winter-Vann, A. (1999) Expression, purification, and structural characterization of the bacteriorhodopsin-aspartyl transcarbamylase fusion protein. *Protein Expr. Purif.* **17**: 324–338.

Ulrich, C.D., Ferber, I., Holicky, E., Hadac, E., Buell, G. and Miller, L.J. (1993) Molecular cloning and functional expression of the human gallbladder cholecystokinin A receptor. *Biochem. Biophys. Res. Commun.* **193**: 204–211.

Unger, V.M., Kumar, N.M., Gilula, N.B. and Yeager, M. (1999) Three-dimensional structure of a recombinant gap junction membrane channel. *Science* **283**: 1176–1180.

van der Laan, M., Bechtluft, P., Kol, S., Nouwen, N. and Driessen, A.J. (2004) F1F0 ATP synthase subunit c is a substrate of the novel YidC pathway for membrane protein biogenesis. *J. Cell Biol.* **165**: 213–222.

van der Vossen, J.M., van der Lelie, D. and Venema, G. (1987) Isolation and characterization of *Streptococcus cremoris* Wg2-specific promoters. *Appl. Environ. Microbiol.* **53**: 2452–2457.

Voet-van-Vormizeele, J. and Groth, G. (2003) High-level expression of the *Arabidopsis thaliana* ethylene receptor protein ETR1 in *Escherichia coli* and purification of the recombinant protein. *Protein Expr. Purif.* **32**: 89–94.

Voss, T., Ergulen, E., Ahorn, H., Kubelka, V., Sugiyama, K., Maurer-Fogy, I. and Glossl, J. (1993) Expression of human interferon omega 1 in Sf9 cells. No evidence for complex-type N-linked glycosylation or sialylation. *Eur. J. Biochem.* **217**: 913–919.

Wagner, S., Bader, M.L., Drew, D. and de Gier, J.W. (2006) Rationalizing membrane protein overexpression. *Trends Biotechnol.* **24**: 364–371.

Wagner, S., Baars, L., Ytterberg, A.J., Klussmeier, A., Wagner, C.S., Nord, O., Nygren, P.A., van Wijk, K.J. and de Gier, J.W. (2007) Consequences of membrane protein overexpression in *Escherichia coli*. *Mol. Cell Proteomics.* **6**: 1527–1550.

Wagner, S., Klepsch, M.M., Schlegel, S., Appel, A., Draheim, R., Tarry, M., Hogbom, M., van Wijk, K.J., Slotboom, D.J., Persson, J.O. and de Gier, J.W. (2008a) Tuning *Escherichia coli* for membrane protein overexpression. *Proc. Natl. Acad. Sci. USA* **105**: 14371–14376.

Wagner, S., Pop, O.I., Haan, G.J., Baars, L., Koningstein, G., Klepsch, M.M., Genevaux, P., Luirink, J. and de Gier, J.W. (2008b) Biogenesis of MalF and the MalFGK(2) maltose transport complex in *Escherichia coli* requires YidC. *J. Biol. Chem.* **283**: 17881–17890.

Walhout, A.J., Temple, G.F., Brasch, M.A., Hartley, J.L., Lorson, M.A., van den Heuvel, S. and Vidal, M. (2000) GATEWAY recombinational cloning: application to the cloning of large numbers of open reading frames or ORFeomes. *Methods Enzymol.* **328**: 575–592.

Walravens, K., Matheise, J.P., Knott, I., Coppe, P., Collard, A., Didembourg, C., Dessy, F., Kettmann, R. and Letesson, J.J. (1996) Immunological response of mice to the bovine respiratory syncytial virus fusion glycoprotein expressed in recombinant baculovirus infected insect cells. *Arch. Virol.* **141**: 2313–2326.

Walter, P., Ibrahimi, I. and Blobel, G. (1981) Translocation of proteins across the endoplasmic reticulum. I. Signal recognition protein (SRP) binds to in-vitro-assembled polysomes synthesizing secretory protein. *J. Cell Biol.* **91**: 545–550.

Warne, T., Serrano-Vega, M.J., Baker, J.G., Moukhametzianov, R., Edwards, P.C., Henderson, R., Leslie, A.G., Tate, C.G. and Schertler, G.F. (2008) Structure of a beta1-adrenergic G-protein-coupled receptor. *Nature* **454**: 486–491.

Weinshank, R.L., Zgombick, J.M., Macchi, M.J., Branchek, T.A. and Hartig, P.R. (1992) Human serotonin 1D receptor is encoded by a subfamily of two distinct genes: 5-HT1D alpha and 5-HT1D beta. *Proc. Natl. Acad. Sci. USA* **89**: 3630–3634.

Weiss, H.M., Haase, W., Michel, H. and Reilander, H. (1995) Expression of functional mouse 5-HT5A serotonin receptor in the methylotrophic yeast *Pichia pastoris*: pharmacological characterization and localization. *FEBS Lett.* **377**: 451–456.

Weiss, H.M., Haase, W. and Reilander, H. (1998) Expression of an integral membrane protein, the 5HT5A receptor. *Methods Mol. Biol.* **103**: 227–239.

White, M.A., Clark, K.M., Grayhack, E.J. and Dumont, M.E. (2007) haracteristics affecting expression and solubilization of yeast membrane proteins. *J. Mol. Biol.* **365**: 621–636.

White, S.H. (2009) Biophysical dissection of membrane proteins. *Nature* **459**: 344–346.

Xia, D., Yu, C.A., Kim, H., Xia, J.Z., Kachurin, A.M., Zhang, L., Yu, L. and Deisenhofer, J. (1997) Crystal structure of the cytochrome bc1 complex from bovine heart mitochondria. *Science* **277**: 60–66.

Zhang, G., Hubalewska, M. and Ignatova, Z. (2009) Transient ribosomal attenuation coordinates protein synthesis and co-translational folding. *Nat. Struct. Mol. Biol.* **16**: 274–280.

Zhang, G. and Ignatova, Z. (2009) Generic algorithm to predict the speed of translational elongation: implications for protein biogenesis. *PLoS. One* **4**: e5036.

Zweers, J.C., Wiegert, T. and van Dijl, J.M. (2009) Stress-responsive systems set specific limits to the overproduction of membrane proteins in *Bacillus subtilis*. *Appl. Environ. Microbiol.* **75**: 7356–7364.

Cold Denaturation and Protein Stability

Piero Andrea Temussi

Abstract Unfolding of mesophilic proteins occurs both at temperatures higher and lower than room temperature: the high temperature transition is generally referred to as "heat denaturation" whereas that at lower temperatures is known as "cold denaturation". We have recently identified a protein, Yfh1, whose cold denaturation occurs at accessible temperatures close to 0°C and under physiological conditions at pH 7; that is, without the need to add denaturants. The first instance in which this system was used in a general sense to study the stability of proteins was a study on the influence of alcohols at low concentrations. Measuring both thermal denaturations, and hence the stability curve, in the presence of trifluoroethanol, ethanol and methanol, we observed an extended temperature range of protein stability. We suggest that alcohols, at low concentration and physiological pH, stabilize proteins by greatly widening the range of temperatures over which the protein is stable. A second important application is illustrated by titin I28, the second case of a protein undergoing unbiased cold denaturation. The thermal stability of this protein cannot be determined by increasing the temperature because aggregation competes with unfolding. The possibility of measuring cold denaturation hints that it is possible to determine accurately thermal stability of many proteins undergoing aggregation.

Keywords Cold denaturation · Misfolding diseases

Abbreviations

CD	Circular dichroism spectroscopy
ΔC_p	Difference in heat capacity between the folded and denaturated states of a protein
ΔG	Difference in free energy between the the folded and denaturated states of a protein

P.A. Temussi (✉)
Dipartimento di Chimica, Universita' di Napoli Federico II, I-80126 Napoli, Italy;
National Institute for Medical Research – MRC, NW7 1AA London, UK
e-mail: temussi@unina.it

J. Brnjas-Kraljević, G. Pifat-Mrzljak (eds.), *Supramolecular Structure and Function 10*, DOI 10.1007/978-94-007-0893-8_4,

ΔH_m	Difference in enthalpy between the folded and denaturated states of a protein
EtOH	Ethanol
HSQC	Heteronuclear single quantum correlated spectrum
MetOH	Methanol
PAG	Polyacrylamide gel
PEG	Polyethylene glycol
T_c	Cold denaturation melting temperature
TFE	2,2,2-trifluoroethanol
T_m	Heat denaturation melting temperature
T_S	Temperature corresponding to the maximum of the stability curve and to the condition $\Delta S = 0$

1 Introduction

Protein unfolding can occur at temperatures both higher and lower than room temperature. Unfolding at higher temperature is generally referred to as "thermal denaturation" whereas unfolding at lower temperature is termed "cold denaturation" (Privalov 1990). Cold denaturation is now a well established concept in protein chemistry. However, it is not uncommon, when referring to it in many scientific communities to have people say: "Cold denaturation? what do you mean?"

Is there something special behind cold denaturation? No, the only problem is essentially psychological: "warm" unfolding is an intuitive concept because it is common experience that heating can degrade most systems; by the same token, "cold" unfolding is counter-intuitive because lowering the temperature in most cases stabilizes systems and slows down processes.

However, the first observation that low temperatures can be detrimental to the stability of proteins is as old as 1930. Hopkins (1930) observed that the rate of denaturation of ovalbumin by urea is higher at 0°C than at 23°C. Many examples of historical interest are reported in the quoted comprehensive review of Privalov (1990). It is easy to rationalize cold denaturation on the basis of the stability curve introduced by Becktel and Schellman (1987). These authors have shown that the difference in free energy between the denatured state and the native state of a protein (ΔG), described by a modified Gibbs-Helmholtz equation, has a convex dependence on temperature, with a maximum at a temperature (T_S) that is generally close to room temperature. Therefore, destabilization of the native state occurs as the temperature varies from room temperature in either direction, when the curve crosses the two zero points of ΔG.

Why is it important to study cold denaturation? Accurate analysis of both the high and low temperature transitions of several proteins could give us a better understanding of protein folding and protein stability. This knowledge has become more and more crucial recently because it has become acutely clear that misfolding, followed by protein aggregation is the main characteristic of several neurodegenerative diseases:"misfolding diseases". From a very general point of view, understanding protein folding is important to explain why some families of proteins show great variability of thermal stability. In addition, there are specific points that, apparently, can only be addressed by low temperature studies. For instance, studies of cold denaturation have already shed light on specific problems otherwise difficult to rationalise. For instance, in the case of the cooperativity of the two domains of phosphoglycerate kinase, Griko et al. (1988) were able to show that cooperativity can be evidenced only at low temperature. The denatured state at low temperature might be intrinsically different from that at high temperature. Evidence in favour of this has been gathered recently in a study of the cold denaturation of ubiquitin in reverse micelles. A marked conformational heterogeneity seems to characterize the ubiquitin ensemble under low-temperature conditions (Whitten et al. 2006). These points and many others have waited very long because, up to very few years ago, it has been impossible to study cold denaturation of a protein in unbiased conditions.

Cold denaturation is difficult to study because it occurs, for most proteins, at temperatures well below freezing of aqueous solutions. To circumvent this difficulty several researchers have tried either to design ways to keep water in a supercooled condition or to raise the temperature of cold denaturation, mainly by destabilizing the protein through mutations and/or denaturants. The obvious drawback of approaches based on artificial denaturation of proteins is the difficulty of extrapolating the results to physiological conditions. We have been seeking new methods to study bio-molecules at subzero temperature and looked for a protein whose cold denaturation could be studied without the help of any destabilization. We have recently identified a protein, Yfh1, whose cold denaturation occurs at accessible temperatures close to 0°C and at physiological conditions (Pastore et al. 2007a). Yfh1 is a small mitochondrial protein and the yeast ortholog of frataxin, the human protein responsible for the neurodegenerative disease Friedreich's ataxia. The frataxin family is highly conserved both in sequence and structure from bacteria to humans and has shown to be essential for life. Despite their conservation, the thermodynamic stability of different orthologs has been shown to vary appreciably. Of the three orthologs so far studied in detail in the literature, Yfh1 is the protein with the lowest and the highest thermal stabilities at high and low temperatures respectively, having the two melting points around 35 and 5°C when in its apo form. These features, together with the fact Yfh1 is a protein from natural sources rather than an ad hoc designed mutant, makes it a system uniquely suited for an extensive characterization of the cold transition and of the factors influencing its stability as a function of temperature. Characterization for this system has allowed investigations on factors that affect thermal denaturations in a new and original way.

2 Unbiased Cold Denaturation

As mentioned above, since the cold denaturation of most proteins occurs at temperatures below the freezing point of water, access to the cold denatured state of most proteins is normally hampered by the obvious limitation of water freezing.

The most used ways to circumvent this difficulty have been to raise the temperature of cold denaturation by destabilising the protein by chemical denaturants, extreme pH values or very high pressure (Kitahara et al. 2006; Griko et al. 1988, Chen and Schellman 1989, Jonas, 2002). In some instances the proteins were destabilized by a combination of point mutations and denaturing agents (Chen and Schellman 1989). The main drawback of these approaches is that it cannot be taken for granted that it is possible to extrapolate the results to meaningful physiological conditions. Even in studies using methods that keep water in a supercooled condition, proteins were destabilized to observe cold denaturation (Babu et al. 2004, Szyperski et al. 2006).

Following a different approach, we looked for a protein whose cold denaturation could be studied without the need for destabilization in a normal buffer at physiological pH within a temperature range accessible to several techniques. In a systematic study of the factors that influence the thermal stability of members of the frataxin family, we had previously shown that although they share the same fold, three orthologues from *E. coli* (CyaY), *S. cerevisiae* (Yfh1) and *H. sapiens* (hfra), are characterized, under the same conditions, by a remarkable variation of melting temperatures (Adinolfi et al. 2002). Yfh1, the one with lowest heat denaturation temperature, seemed a promising candidate to observe cold denaturation above the freezing point of aqueous solutions.

Both NMR and CD experiments showed that Yfh1 can undergo cold denaturation at temperatures above the freezing point of water in unbiased solution conditions.

Figure 1 (top panel) shows the plot of the integrals of four well separated high field peaks as a function of temperature. The points fall on a curve that indicates a

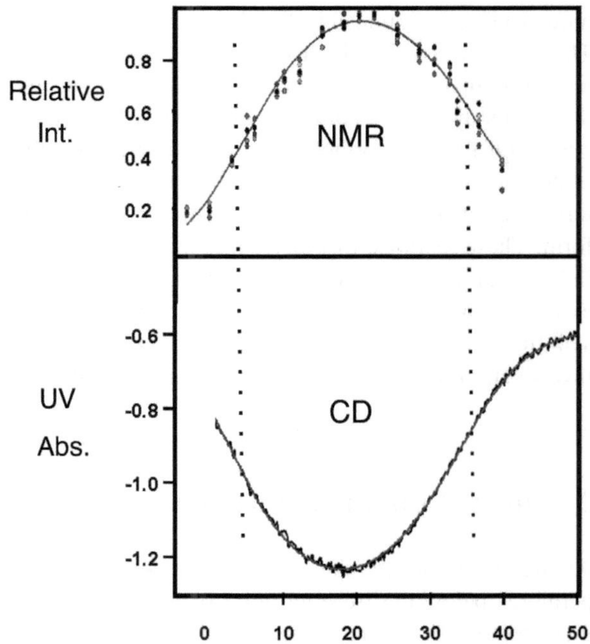

Fig. 1 Plots of NMR and CD signals of Yfh1 as a function of temperature. Relative integrals of the highest field peaks of Yfh1 in HEPES at pH 7.0 are shown in the *top panel*. Far UV absorbance in the CD spectra of Yfh1 in 20 mM Tris at pH 7.5 are plotted in the *bottom panel*

low temperature transition at ca. 7°C and a high temperature transition at ca. 31°C. CD measurements, shown in the bottom panel of Fig. 1, are consistent with the same transition temperatures (Pastore et al. 2007a). Analysis of the curves allowed the measurement of all relevant thermodynamic parameters for both transitions, revealing that the possibility of observing them above the freezing temperature of water is linked to the exceptionally high value of ΔCp (7.57 kJ/mol). This work showed that Yfh1 is an excellent model system to study low and high temperature transitions in a range accessible to many techniques, and indeed it was possible to exploit it to unveil the action of alcohols on protein stability.

3 Alcohols: Denaturants or Structure Stabilizers?

There were hints in the literature that alcohols, although universally considered well known protein denaturants at high concentrations, could have a different effect on proteins at low concentrations (Brandts and Hunt 1967, Povey et al. 2007). Exploiting the unusual property of Yfh1 of undergoing cold denaturation around 0°C without any ad hoc destabilization, Martin et al. (2008) were able to determine the stability curve this protein on the basis of both high and low temperature unfolding in the presence of three common alcohols: trifluoroethanol, ethanol and methanol. In a general case, i.e. when it is necessary to destabilize the protein to have access to cold denaturation, it would be impossible to observe the influence of yet another potential destabilizing agent, such as an alcohol.

Yfh1 seemed particularly well suited for studies of cold denaturation in water-alcohol mixtures because, starting from a cold denaturation melting temperature (T_c) already above zero, it should still be possible to observe a substantial part of the denaturation curve even if addition of alcohols shifts T_c to temperatures lower than 0°C.

Figure 2 shows the thermal denaturation curves of Yfh1 measured by monitoring the CD intensity at 222 nm as a function of temperature in the temperature range 0–60°C for two of the water-alcohol mixtures: methanol (top panel) and trifluoroethanol (bottom panel). It is possible to observe that even very small percentages of alcohol have a profound effect on the form of the denaturation curve: the high temperature unfoldings (T_m) are not much affected, whereas the low temperature transitions (T_c) are shifted to even lower temperatures, thus widening the range of temperatures over which the protein is stable.

When changes in protein conformation were monitored by CD and NMR spectrocopies Martin et al. (2008) observed that small amounts of alcohols lead to the unusual effect of decreasing ΔC_p, leading to a flattening of the stability curve of the protein. In addition, a concomitant increase of ΔH_m leads to a decrease of T_S, the temperature of maximum stability, thus shifting the whole stability curve to lower temperatures and widening the temperature range over which the protein is stable.

As a consequence of the increase of the ratio $\Delta H_m/\Delta C_p$ we may have a partial compensation of the increase of T_m, and an enhancement of the decrease of T_c.

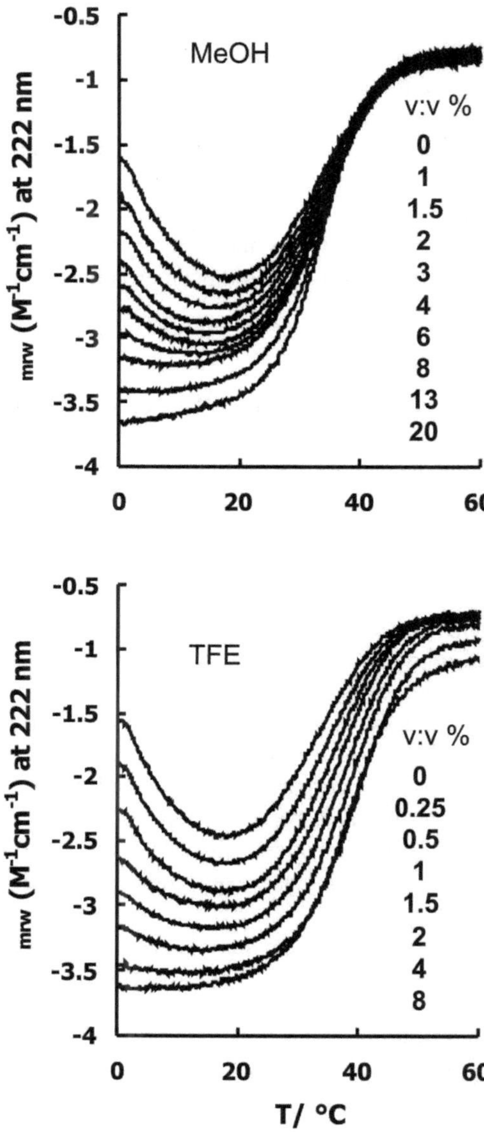

Fig. 2 Thermal denaturation curves of Yfh1 in a 20 mM HEPES buffer at pH 7.0 with a protein concentration of 10 μM, measured by monitoring the CD intensity at 222 nm as a function of temperature in the temperature range 0–60°C in two of the buffer-alcohol mixtures. Spectra comprised between the two extremes have the molar concentrations indicated on the *right*. *Top panel* shows the denaturation curves in methanol-water. *Bottom panel* shows the denaturation curves in trifluoroethanol-water

The great advantage offered by Yfh1 resides in the possibility to observe directly a large part of the cold denaturation curve. In other cases, with a slightly different combination of values for ΔH_m and ΔC_p, the shift of the stability curve towards lower temperatures can be more pronounced, leading to a moderate decrease of T_m as observed by Velicelebi and Sturtevant (1979) for lysozyme and by Fu and Freire (1992) for cytochrome c. Thus, it is possible to have a concomitant widening of the temperature range of protein stability and a decrease of T_m. A graphic illustration of these possibilities, based on the scheme proposed by Nojima et al.

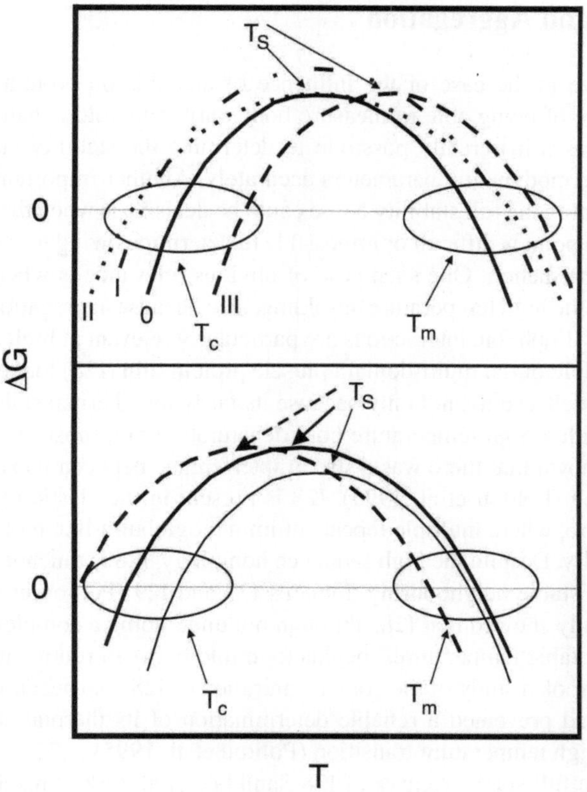

Fig. 3 Influence of thermodynamic parameters on stability curves of a protein, adapted from Nojima et al. (1977). *Top panel*: temperature dependence of the difference of free energy between denatured and native states (ΔG) of the original protein (curve 0, *thin black line*) and the effect of three different limiting mechanisms. Mechanism I shows the effect of increasing the value of ΔH_S (*curve I, .-.- line*). Mechanism II shows the consequence of reducing the value of vC_p (*curve II, ... line*). Mechanism III, a shows the shift of the whole stability curve towards higher temperatures caused by a lowering of ΔS_m (*curve III, --- line*). *Bottom panel*: two different combinations of the three mechanisms. The *solid thick curve* corresponds qualitatively to the cases of Yfh1 reported by Martin et al. (2008). The *dashed curve* corresponds qualitatively to the cases reported by Velicelebi and Sturtevant (1979) and by Fu and Freire (1992)

(1977) to rationalise the stability of proteins from thermophilic organisms is reported in Fig. 3.

Thus, Martin et al. (2008) were able to generalize the effect of alcohols, observing that they, at low concentration and physiological pH, stabilize proteins by widening the range of temperatures over which the protein is thermally stable. A direct experimental observation of this stabilization requires an accurate determination of the whole stability curve. This is actually possible with a system as Yfh1, but it may be rather difficult in a general case, owing to the inaccessibility of cold denaturation temperatures.

4 Stability and Aggregation

As stated above in the case of the influence of alcohols on protein stability, the main advantage of being able to measure both warm and cold denaturation is that only in such cases it is really possibile to determine the stability curve and thus measure all thermodynamic parameters accurately. Another important instance for which determining the full stability curve could be decisive is when the high temperature transition point is difficult or impossible to determine owing to the competition of another phenomenon. One such case of obvious relevance is when aggregation interferes with the high temperature unfolding, also because aggregation phenomena governed by hydrophobic interactions are particularly relevant at high temperatures. A domain module of the multi-domain muscle protein titin (I28) looked as the ideal candidate for such a study, not only because its fairly low thermal stability could be compatibile with a high temperature cold denaturation but, most of all, because it was already known that there was a strong interference between thermal unfolding and aggregation (Politou et al. 1995). I28 is present in the elastic region (I-band) of the sarcomere, where multiple repeats of immunoglobulin-like motifs are assembled sequentially. Despite the high sequence homology, I28 is much less stable than its remarkably stable neighbouring domains I27 and I29 (Politou et al. 1995). An exploratory study showed that I28, although not undergoing a complete cold denaturation at detectable temperatures, begins to unfold between room temperature and 0°C. We undertook a study of the cold denaturation of I28 also because its tendency to aggregate had prevented a reliable determination of its thermal stability based solely on the high temperature transition (Politou et al. 1995).

To study the full stability curve of I28 Sanfelice et al. (2009) resorted to NMR, designing conditions that can influence aggregation significantly and, at the same time, facilitate the attainement of very low temperatures, in case the cold denaturation temperature were well below 0°C. This novel approach that exploits the combined use of polyethylene glycol (PEG) and polyacrylamide gel (PAG), thus combining confinement with crowding, is similar to the use of bundles of capillaries successfully employed in the past (Szyperski et al. 2006) but has the advantage of overcoming some of the difficulties intrinsic in handling capillaries, such as the risk of breakages of the micro-tubes. The internal structure of the gel may be thought of as a collection of thin capillaries, and the irregular structure of the cavities may favour attainment of low temperatures without risking crystallization in some of the capillaries (Pastore et al. 2007b). By comparing the unfolding process in bulk and in the gel, Sanfelice et al. (2009) demonstrated that this approach allows the characterization of the entire curve of stability of I28 and provides useful and novel information about denaturation in crowded and confined environments.

A comparison of ^{15}N-HSQC NMR spectra of titin I28 at different temperatures in the range from –15 to 60°C revealed that both raising and lowering the temperature with respect to room temperature causes a uniform decrease in the volumes of the peaks, consistent with an unfolding transition. However, the two transitions are not equally reversible. When starting from the lowest attainable temperature, the ensuing heating process does restore the full intensity of room temperature signals indicating a full recovery of the initial fold. On the contrary, heating the sample to temperatures higher than 328 K leads to precipitation and subsequent cooling back to room temperature does not restore the folded state.

In order to extend the temperature range to lower subzero values Sanfelice et al. (2009) resorted to the environment created inside the cavities of PAGs of different mesh by the addition of solutions of PEG of several molecular weights. The corresponding spectra at room temperature, in bulk and in gel are very similar, confirming that it is possible to use the confined environment to study unfolding processes, starting from the natural fold. However, the simultaneous presence of both PAG and PEG determines crowding and confinement conditions that further favor aggregation of the protein. This is hardly surprising because confining environments are known to greatly favour aggregation phenomena more than the stabilization of compact, folded structures (Minton 2000).

NMR data, both in bulk and in the crowded environment, could be used for a direct evaluation of the thermodynamic parameters of both cold and heat denaturation of I28. Figure 4 shows that by monitoring the volumes of four isolated NH resonances it is possible to follow the thermal stability over the entire range,

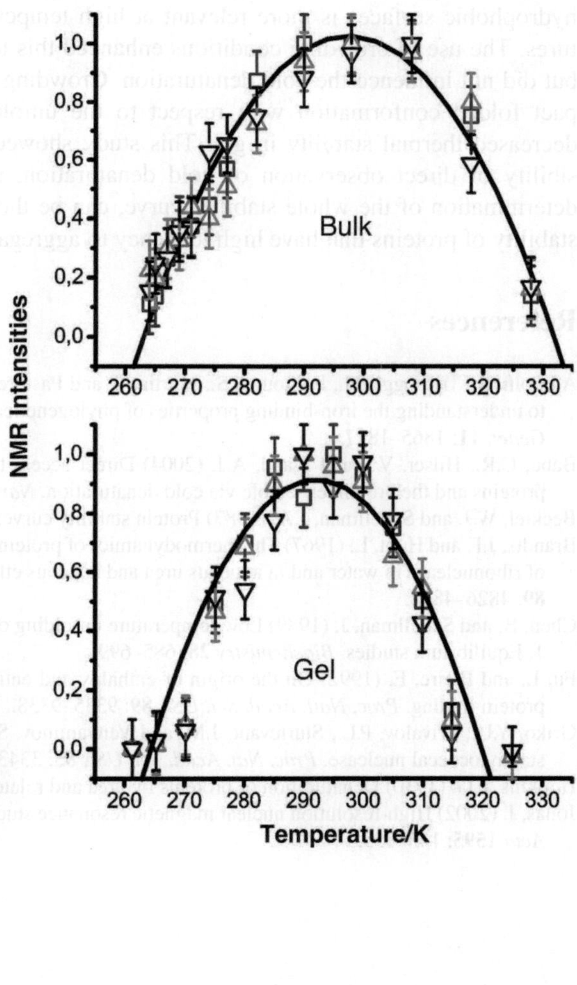

Fig. 4 Quantitative evaluation of NMR data of I28. Variation of the peak volumes of four I28 resonances as a function of temperature in bulk 25 mM TRIS pH 7.2 (*top panel*). Variation of the peak volumes as a function of temperature in 25 mM TRIS, 5% PAG and 10% PEG2000 at pH 7.2 (*bottom panel*)

Table 1 Comparison of the thermodynamic parameters of I28 in bulk and in gel

	Bulk	Gel
T_m/K	318	310
$\Delta H_m°$/ kcal mol^{-1}	38.2	46.5
$\Delta S_m°$/ cal mol^{-1}K^{-1}	120	150
ΔCp/ kcal mol^{-1}K^{-1}	1.77	2.7
T_c/K	275	275
$\Delta H_c°$/ kcal mol^{-1}	–38.2	–46.5
$\Delta S_c°$/ cal mol^{-1}K^{-1}	120	150

covering both the high and low temperature unfolding. Fitting the curves with the Becktel–Schellman equation (1987), Sanfelice et al. (2009) calculated the thermodynamic parameters reported in Table 1. The most interesting aspect of these data is the fact that, whereas the low transition temperatures coincide for bulk solution and gel, the high temperature transitions are appreciably different.

The natural interpretation of this discrepancy is that the confined environment accelerates aggregation at high temperatures and thus the lower T_m reflects the competition of two phenomena: unfolding and aggregation. Aggregation involving hydrophobic surfaces is more relevant at high temperatures than at low temperatures. The use of crowding conditions enhanced this tendency at high temperature but did not influence the cold denaturation. Crowding should favor the more compact folded conformation with respect to the unfolded one, but we observe a decreased thermal stability in gel. This study showed convincingly that the possibility of direct observation of cold denaturation, and the consequent reliable determination of the whole stability curve, can be the only way to assess thermal stability of proteins that have high tendency to aggregate at high temperatures.

References

Adinolfi, S., Trifuoggi, M., Politou, A.S., Martin, S. and Pastore, A. (2002) A structural approach to understanding the iron-binding properties of phylogenetically different frataxins. *Hum. Mol. Genet.* **11**: 1865–1877.

Babu, C.R., Hilser, V.J. and Wand, A.J. (2004) Direct access to the cooperative substructure of proteins and the protein ensemble via cold denaturation. *Nat. Struct. Mol. Biol.* **11**: 352–357.

Becktel, W.J. and Schellman, J.A. (1987) Protein stability curves. *Biopolymers* **26**: 1859–1877.

Brandts, J.F. and Hunt, L. (1967) The thermodynamics of protein denaturation. 3. The denaturation of ribonuclease in water and in aqueous urea and aqueous ethanol mixtures. *J. Am. Chem. Soc.* **89**: 4826–4838.

Chen, B. and Schellman, J. (1989) Low-temperature unfolding of a mutant of phage T4 lysozyme. 1. Equilibrium studies. *Biochemistry* **28**: 685–699.

Fu, L. and Freire, E. (1992) On the origin of enthalpy and entropy convergence temperatures in protein folding. *Proc. Natl. Acad. Sci. USA* **89**: 9335–9338.

Griko, Y.U., Privalov, P.L., Sturtevant, J.M. and Venyaminov, S.Yu. (1988) Cold denaturation of staphylococcal nuclease. *Proc. Nat. Acad. Sci. USA* **85**: 3343–3347.

Hopkins, F.G. (1930) Denaturation of proteins by urea and related substances. *Nature* **126**: 383.

Jonas, J. (2002) High-resolution nuclear magnetic resonance studies of proteins. *Biochim. Biophys. Acta* **1595**: 145–159.

Kitahara, R., Okuno, A., Kato, M., Taniguchi, Y., Yokoyama, S. and Akasaka, K. (2006) Cold denaturation of ubiquitin at high pressure. *Magn. Reson. Chem.* **44**: S108–S113.

Martin, S.R., Esposito, V., De Los Rios, P., Pastore, A. and Temussi, P.A. (2008) The effect of low concentrations of alcohols on protein stability: a cold and heat denaturation study of yeast frataxin *J. Am. Chem. Soc.* **130**: 9963–9970.

Minton, A.P. (2000) Effect of a concentrated "inert" macromolecular cosolute on the stability of a globular protein with respect todenaturation by heat and by chaotropes: a statistical-thermodynamic model. *Biophys. J.* **78**: 101–109.

Nojima, H., Ikai, A., Oshima, T. and Noda, H. (1977) Reversible thermal unfolding of thermostable phosphoglycerate kinase. Thermostability associated with mean zero enthalpy change. *J. Mol. Biol.* **116**: 429–442.

Pastore, A., Martin, S.R., Politou, A., Kondapalli, K.C., Stemmler, T. and Temussi, P.A. (2007a) Unbiased cold denaturation: low- and high-temperature unfolding of yeast frataxin under physiological conditions. *J. Am. Chem. Soc.* **29**: 5374–5375.

Pastore, A., Salvadori, S. and Temussi, P.A. (2007b) Peptides and proteins in a confined environment: NMR spectra at natural isotopic abundance. *J. Pept. Sci.* **13**: 342–347.

Politou, A.S., Thomas, D.J. and Pastore, A. (1995) The folding stability of titin immunoglobin-like modules, with implications fort he mechanism of elasticity. *Biophys. J.* **69**: 2601–2610.

Povey, J.P., Smales, C.M., Hassard, S.J. and Howard, M.J. (2007) Comparison of the effects of 2,2,2-trifluoroethanol on peptide and protein structure and function. *J. Struct. Biol.* **157**: 329–338.

Privalov, P.L. (1990) Cold denaturation of proteins. *Crit. ReV. Biochem. Mol. Biol.* **25**: 281–305.

Sanfelice, D., Tancredi, T., Politou, A., Pastore, A. and Temussi, P.A. (2009) Cold denaturation and aggregation: a comparative NMR study of titin I28 in bulk and in a confined environment. *J. Am. Chem. Soc.* **131**: 11662–11663.

Szyperski, T., Mills, J.L., Perl, D. and Balbach, J. (2006) Combined NMR-observation of cold denaturation in supercooled water and heat denaturation enables accurate measurement of deltaC(p) of protein unfolding. *Eur. Biophys. J.* **35**: 363–366.

Velicelebi, G. and Sturtevant, J.M. (1979) Thermodynamics of the denaturation of lysozyme in alcohol-water mixtures. *Biochemistry* **18**: 1180–1186.

Whitten, S.T., Kurtz, A.J., Pometun, M.S., Wand, A.J. and Hilser, V.J. (2006) Revealing the nature of the native state ensemble through cold denaturation. *Biochemistry* **45**: 10163–10174.

Polyglutamine Diseases and Neurodegeneration: The Example of Ataxin-1

Cesira de Chiara and Annalisa Pastore

Abstract A family of nine human neurodegenerative diseases is caused by anomalous expansion of polyglutamine (polyQ) tracts in the carrier proteins. Understanding the cellular and molecular mechanisms which lead to disease is essential both for understanding these pathologies and for designing appropriate diagnostics. We review here, as a paradigmatic example, the knowledge accumulated for ataxin-1, the protein responsible for Spinocerebellar Ataxia type 1 (SCA1) and one of the smallest representatives of the polyQ family. It appears clear from this overview that understanding the properties and the interaction networks formed by both expanded and non-expanded ataxin-1 is an essential step for our comprehension of the non-pathologic function of the protein and of its role in disease. A better understanding will be reached in the future from integrating knowledge arising from different fields.

Keywords Ataxin-1 · SCA1 · Neurodegenerative disease · Polyglutamine · CAG triplet · AXH · ULM · Molecular switch · Transcriptional repression · RNA processing · Protein context

1 Introduction

Ataxin-1 (Atx1) is a 98 kDa protein of yet unknown function responsible for the human disease spinocerebellar ataxia type 1 (SCA1), an autosomal dominant neurodegenerative disorder characterized by motor coordination deficits caused by progressive loss of Purkinje cells in the cerebellar cortex, and neurons in the brain stem and spinocerebellar tracts (Banfi et al. 1994, Jodice et al. 1994, Orr and Zoghbi 2001). SCA1 is a member of a clinically and genetically heterogeneous group of autosomal dominant diseases, which include Huntington's chorea, Machado–Joseph disease and different other spinocerebellar ataxias (Orr and Zoghbi 2007). From a

---

A. Pastore (✉)
National Institute for Medical Research – MRC, NW7 1AA, London, UK
e-mail: apastor@nimr.mrc.ac.uk

J. Brnjas-Kraljević, G. Pifat-Mrzljak (eds.), *Supramolecular Structure and Function 10*, DOI 10.1007/978-94-007-0893-8_5,

clinical point of view, most of these diseases are characterized by loss of balance and motor coordination caused by cerebellum dysfunction (Zoghbi 2000, Zoghbi and Orr 2000). Typical clinical features of SCA1 in patients include gait ataxia, dysarthria, and bulbar dysfunction, with death usually between 10 and 15 years after the onset of symptoms (Matilla-Duenas et al. 2008).

At the genetic level, SCA1, like the other members of the family, is caused by expansion of CAG trinucleotide repeats coding for a polymorphic polyglutamine (polyQ) tract in the gene products (Banfi et al. 1994, Jodice et al. 1994, Orr and Zoghbi 2001, Zoghbi and Orr 2009). The disorder typically develops when the repeat length exceeds a threshold of 35–45 glutamines (Genis et al. 1995, Jayaraman et al. 2009, Orr et al. 1993). PolyQ expansions induce protein aggregation and formation of neuronal intranuclear inclusions, a hallmark of the polyQ diseases and consequent cell death (Klement et al. 1998, 1999). Intranuclear aggregates contain expanded and non-expanded polyQ proteins and several other cellular components, such as ubiquitin, molecular chaperones, members of the proteasomal apparatus, and transcription factors (Bauer and Nukina 2009, Shao and Diamond 2007).

While it is now widely accepted that the pathologic symptoms are caused by the toxic effects of expanded polyQ tracts in Atx1 and in the other members of the family, the molecular and cellular events underlying neurodegeneration are still unclear. Recent evidence has indicated for instance that also non-expanded polyQ proteins tend to aggregate and that protein context has a strong effect on protein solubility and stability (de Chiara et al. 2005b; Masino et al. 2004, Saunders and Bottomley 2009). Other regions of polyQ proteins could therefore be at least partly responsible for the formation of inclusions and be involved in pathogenesis.

Here, we review our knowledge on the functions of Atx1 and show how different approaches can add valuable information for understanding pathology.

2 Structural Features of Atx1

2.1 Atx1 Domain Architecture

Atx1 is a ca. 800 residues protein, depending on polyQ length (Fig. 1a). The polyQ tract is at residue 197 (human sequence), that is close to the N-terminus of the protein. Outside this region, most of its sequence contains low complexity and/or potentially unfolded regions, suggesting that the molecule is mostly a constitutively unfolded protein. The only exception is the AXH motif (SMART database Acc. Name SM00536) (Mushegian et al. 1997), present near the C-terminus, which is currently the only known folded region of the molecule. The AXH module was first identified as a sequence motif through its homology with a region of the HMG box-containing transcription factor 1 HBP1, another otherwise unrelated protein (Lesage et al. 1994). While clearly correlated (the two sequences share *ca.* 30% identity and *ca.* 50% homology, depending on the species), the AXH motif has slightly different domain boundaries and distinct properties in the two protein subfamilies (de Chiara et al. 2003, 2005a, b).

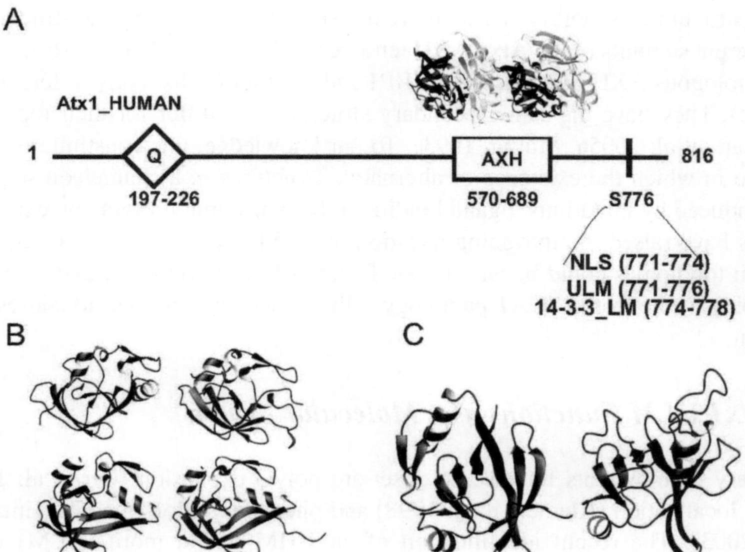

Fig. 1 Structure of Atx1. (**a**) Schematic representation of the domain architecture of human Atx1. Q stands for the polyQ tract. A ribbon representation of the three dimensional structure of the AXH domain of Atx1 (PDB entry code 1oa8) is shown on the AXH module. The different subunits of the dimer of dimers are shown in alternate black and gray colour. (**b**) Comparison of the four subunits in 1oa8 shown in the same orientation. (**c**) Comparison of the structure of the AXH motif of Atx1 (*left*) and HBP1 (*right*)

Three overlapping linear motifs were identified downstream to AXH: a UHM ligand motif (ULM) (de Chiara et al. 2009), present in proteins associated with splicing, a nuclear localization signal (Klement et al. 1998) and a 14-3-3 binding motif (Chen et al. 2003). ULM comprises S776, a residue which can be phosphorylated (Jorgensen et al. 2009). Phosphorylation of S776 is, in addition to polyQ expansion and nuclear localization, a necessary condition for development of SCA1 (Emamian et al. 2003).

2.2 The Structure of the Atx1 AXH Domain

The structure of Atx1 AXH was solved by X-ray crystallography (Chen et al. 2004) (Fig. 1a, b). It consists of a non-canonical oligonucleotide and oligosaccharide-binding (OB) fold (Murzin 1993), which forms a dimer of asymmetric dimers. The Atx1 AXH monomer has a sickle-like shape with the concave and convex sides engaged in the interfaces between each dimer and between the two dimers, respectively. The four subunits have distinct structures and overlap well with each other only when superposing the backbone atoms of residues 610–685 (average root mean square deviation of 0.90 ± 0.06 Å). Outside this region, there are appreciable differences which are mostly localized at the N- and C-termini. Larger variability is observed in regions directly involved in intermolecular interfaces.

The structural variability is even more marked when comparing the structures of the different subunits of the Atx1 AXH tetramer with that of HBP1: the structures of the homologous AXH domains from HBP1 and Atx1 adopt distinctly different folds (Fig. 1c). They have the same secondary structures but differ for their topologies (de Chiara et al. 2005a, Murzin 1993). To our knowledge, this constitutes the first example in which the existence of alternative structures in a chameleon sequence is not induced by mutations, ligand binding or by a different protein context. These findings have raised the intriguing question of whether the possibility of adopting different topologies could be an intrinsic feature of the AXH motif and whether it could play a role in the SCA1 pathology. More research is needed to address this question.

2.3 Atx1 ULM Functions as a Molecular Switch

Necessary requirements for disease onset are polyQ expansion (Orr et al. 1993), nuclear localization (Klement et al. 1998) and phosphorylation of S776 (Emamian et al. 2003). The recent identification of an UHM ligand motif (ULM) in the C-terminus of the Atx1 sequence provided an explanation for these observations. ULMs are present in proteins associated with splicing suggesting a link between Atx1 and RNA processing. It was shown that Atx1 interacts with and influences the function of the splicing factor U2AF65 via this motif (de Chiara et al. 2009). ULM comprises S776 of Atx1 and overlaps with a nuclear localization signal and a 14-3-3 binding motif. Phosphorylation of S776 provides the switch which discriminates binding to the 14-3-3 protein and components of the spliceosome. It is of interest to notice that an S776D Atx1 mutant previously designed to mimic phosphorylation (Lim et al. 2008) is unsuitable for this aim because of the very different chemical properties of the two groups. These results indicate that Atx1 is part of a complex network of interactions with splicing factors, which may play a protective role against aggregation, and suggest that development of the pathology is the consequence of a competition of aggregation with native interactions (Fig. 2). An important and more general corollary of these results is that studies of the interactions formed by non-expanded Atx1 may provide valuable hints not only for understanding the function of the non-pathologic protein but also the causes of disease.

Interestingly, ULM is not conserved in the Atx1 paralogue Boat (Brother of Atx1), which also lacks the polyQ tract and is able to suppress cytotoxicity of mutant Atx1 (Mizutani et al. 2005).

3 Atx1 Functions

3.1 A Role for Atx1 in Transcriptional Repression

Although the exact role of non-pathological Atx1 is still unclear, several findings suggest that it is a transcription factor (Tsai et al. 2004). Indication that the

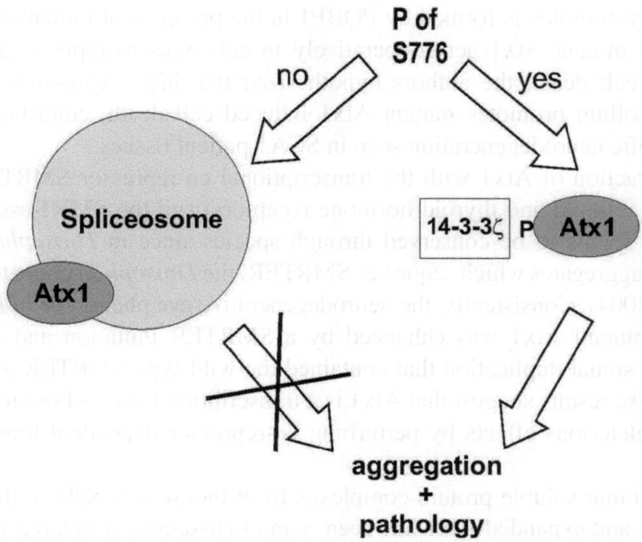

Fig. 2 Diagram of the role of phosphorylation of S776 as a molecular switch. When S776 is not phosphorylated Atx1 participates to the large mesh of interactions formed by spliceosomal factors and is protected against aggregation. When it is phosphorylated, it interacts with 14-3-3 and other such proteins and remains more prone to aggregation

SCA1 pathogenesis is due to alterations in gene expression regulation began with the observation that, to cause disease, expanded Atx1 had to enter the nucleus of Purkinje cells (Klement et al. 1998) and that alterations in gene expression occur very early in SCA1 transgenic mice, prior to the onset of any detectable neurological or pathological changes (Lin et al. 2000, Serra et al. 2004).

More solid evidence in favor of a link between Atx1 function, SCA1 pathogenesis and transcriptional regulation came from a genetic screening in *Drosophila* (Fernandez-Funez et al. 2000) where several transcriptional co-repressors, including Sin3 and Rpd3 (the *Drosophila* histone deacetylase 1), alongside with components involved in protein folding, protein clearance and RNA processing, were identified as modulators of the Atx1 mediated eye phenotype.

Since then, several reports have suggested that Atx1 interacts with and/or modulates the function of several other transcriptional co-regulators. Among them are the polyQ-binding protein-1 PQBP1 (Okazawa et al. 2002), the silencing mediator of retinoid and thyroid hormone receptors SMRT/SMRTER (Tsai et al. 2004), the repressor Capicua (Lam et al. 2006), the transcription factors Sensless/Gfi-1 (Tsuda et al. 2005) and Sp1 (Goold et al. 2007), and the transcriptional complexes Tip60-RORα (Serra et al. 2006) and HDAC-MEF2 (Bolger et al. 2007).

PQBP1 has also been suggested to be involved in the pathology of SCA1 (Okazawa et al. 2002). Using in vitro and in vivo assays, it was shown that the interaction between Atx1 and PQBP1 was influenced by expanded polyQ sequences. In immunoprecipitation experiments, mutant Atx1 enhanced the interaction between PQBP1 and the RNA polymerase II (pol II) large subunit. The authors proposed that a ternary complex is formed by PQBP1 in the presence of mutant Atx1. Since PQBP1 and mutant Atx1 act cooperatively in cell lines to repress transcription and induce cell death, the authors hypothesized that high expression of PQBP1 in the cerebellum promotes mutant Atx1 induced cell death, contributing to the region-specific neurodegeneration seen in SCA1 patient tissues.

The interaction of Atx1 with the transcriptional co-repressor SMRT (silencing mediator of retinoid and thyroid hormone receptors) and the SMRT-associate factor HDAC3 seems to be conserved through species since in *Drosophila*, mutant Atx1 forms aggregates which sequester SMRTER, the *Drosophila* cognate of SMRT (Tsai et al. 2004). Consistently, the neurodegenerative eye phenotype in *Drosophila* caused by mutant Atx1 was enhanced by a SMRTER mutation and suppressed by a chromosomal duplication that contained the wild-type SMRTER gene. Taken together, these results suggest that Atx1 is a transcription factor whose mutant form exerts its deleterious effects by perturbing co-repressor dependent transcriptional pathways.

By examining soluble protein complexes from mouse cerebellum, the majority of wild-type and expanded Atx1 has been found to assemble into large stable complexes containing the transcriptional repressor Capicua (CIC) (Lam et al. 2006) and to modulate its repressor activity in *Drosophila* and mammalian cells. Among all the Atx1 interacting proteins and genetic modifiers of SCA1 phenotypes that are involved in transcriptional regulation, CIC forms a stable rather than transient interaction with Atx1. Interestingly, the S776A mutation, which abrogates neurotoxicity of expanded Atx1, substantially reduces the association of mutant Atx1 with CIC in vivo indicating that incorporation of expanded Atx1 in the large CIC complex is important for toxicity. Concomitantly, polyQ expansion attenuates the formation and the function of the Atx1/CIC-containing complex (Lim et al. 2008). Elevated Boat levels were found to suppress neuropathology by displacing mutant Atx1 from its native complex with CIC (Bowman et al. 2007). These results provide genetic evidence that the selective neuropathology of SCA1 arises from modulation of a core functional activity of Atx1. Together with previous observations (Mizutani et al. 2005), these studies underscored the importance of studying the paralogues of genes mutated in neurodegenerative diseases to gain insight into the mechanisms of pathogenesis.

Other important transcription factors which have been shown to be physical and/or genetic interactors of Atx1 are the *Drosophila* zinc-finger transcription factor Senseless (Sens) and its mammalian homologue growth factor independence-1 (Gfi-1) (Tsuda et al. 2005). The level of Sens and Gfi-1 are reduced both in the fly and mouse models, when wild-type or mutated Atx1 are overexpressed, respectively. The interaction is mediated by the conserved Atx1 AXH domain, since deletion of the AXH domain abolishes the effects of glutamine-expanded mammalian Atx1 on Sensless/Gfi-1.

Atx1 has been shown to interact and function synergistically with the zinc-finger transcription factor Sp1 to co-regulate dopamine receptor D2 (Drd2) expression (Goold et al. 2007). Alterations in expression of genes regulated by Sp1-dependent transcription, including dopamine receptor D2 (Drd2), retinoic acid/thyroid hormone, and Wnt-signalling, have been identified by microarray analysis in Atx1-null mice, thus providing novel molecular targets regulated by Atx1.

Atx1 has been found in complex with RORα, a transcription factor critical for cerebellar development, and with the RORα-coactivator Tip60 (Serra et al. 2006). It was hypothesized that mutant Atx1 may disrupt the complex and destabilize RORα, leading to depletion and reduction in expression of genes controlled by RORα. The authors suggested that a decrease of RORα-mediated gene expression during development is an important factor in the susceptibility of Purkinje cells to SCA1.

The MEF2-HDAC4 transcriptional complex involved in neuron survival was also identified as a target of Atx1 transcriptional repression activity (Bolger et al. 2007). Atx1 binds HDAC4 and MEF2 specifically and colocalizes with them in nuclear inclusion bodies. Significantly, these interactions are greatly reduced by the S776A mutation. The results suggest a novel pathogenic mechanism whereby Atx1 sequesters and inhibits the neuronal survival factor MEF2.

Finally, it should be mentioned the binding to the leucine-rich acidic nuclear protein (LANP). Initially identified as an Atx1-binding protein on the base of a yeast-two-hybrid screen assay, LANP was suggested to be a candidate mediator of neurotoxicity in SCA1 because of its proposed role in cerebellar morphogenesis (Matilla et al. 1997). More recently, it has been shown that Atx1 relieves the transcriptional repression induced by the complex between LANP and the transcriptional repressor E4F, thus establishing the first functional link between LANP and Atx1 (Cvetanovic et al. 2007).

3.2 RNA Processing

Atx1 was shown to bind RNA by in vitro RNA-binding assays. Binding diminishes as the length of the polyQ tract increases (Yue et al. 2001). These findings allowed the authors of the report to hypothesize for the first time that Atx1 might play a role in RNA metabolism.

In support to this hypothesis, it was observed that Atx1 co-localizes in nuclear bodies with the mRNA export factor TAP/NXF1 in a manner that is dependent on the presence of RNA and that wild-type, but not a polyQ expanded mutant, has the ability to export the protein from the nucleus (Irwin et al. 2005). These results suggest that the normal role of Atx1 covers RNA processing and possibly nuclear RNA export and that the nuclear retention of mutant Atx1 may be an important toxic gain of function in SCA1 disease. Along these lines, it was demonstrated that RNA co-localizes with inclusions of expanded Atx1 (de Chiara et al. 2005b) also in absence of the RNA binding AXH domain, thus suggesting the involvement of other RNA binding proteins.

Identification, in the C-terminus of Atx-1, of the ULM motif, which mediates Atx1 interaction with the UHM domains of splicing factors, opens new perspectives on a possible role of Atx1 in pre-mRNA processing (de Chiara et al. 2009, Lim et al. 2008). In support to this hypothesis is also the presence of other RNA binding proteins in the Atx1 interactome (Lim et al. 2006).

3.3 The Multiple Role of the AXH Motif in Nucleic Acid– and Protein–Protein Interactions

AXH is the region responsible for the transcriptional repression activity of Atx1: the isolated AXH domain is able to act as a transcriptional repressor in a general read-out assay for repression of transcription (de Chiara et al. 2005b). In agreement with what observed for full-length Atx1 (Tsai et al. 2004), the AXH domain represses transcription when tethered to DNA. DNA binding was, however, shown to be indirect by cross-linking experiments in cell and, therefore, mediated by other co-transcriptional regulators (de Chiara et al. 2005b).

With the exception of PQBP1 (Okazawa et al. 2002), which binds the polyQ tract, and RORα-Tip60 (Serra et al. 2006) and HDAC-MEF2 complexes (Bolger et al. 2007), whose binding sites have not been mapped in details, all the other interactions with transcriptional regulators mentioned above have been shown to involve the Atx1 AXH domain: deletion of the AXH domain was proved to abolish the interaction of Atx1 with SMRT, Gfi-1, CIC, and Sp1 (Goold et al. 2007, Lam et al. 2006, Mizutani et al. 2005, Tsuda et al. 2005). Thus, this motif has an essential role in the non-pathologic Atx1 functions and, possibly, in SCA1 pathology.

Atx1 was shown to be SUMOylated on at least five lysine residues, two of which are in the AXH domain (Riley et al. 2005). SUMOylation depends on the length of the polyQ tract, the ability of Atx1 to be phosphorylated at Ser776, and the integrity of the nuclear localization signal. The phosphorylation- and nuclear localization-dependence of Atx1 SUMOylation is reminiscent of other nuclear body proteins, namely, PML, Sp100, and HDAC4, in which SUMOylation controls the nucleocytoplasmic trafficking as well as their ability to function as transcriptional corepressors (Johnson 2004). Most targets of SUMOylation are nuclear proteins, many of them having a role in gene transcription (Seeler and Dejean 2003). These findings suggest that Atx1 SUMOylation could be linked to its ability to regulate transcription.

In addition to the role in mediating protein–protein interaction with transcription factors, the AXH domain also binds nucleic acid by virtue of its oligonucleotide binding fold: AXH is sufficient to bind RNA homopolimers with the same nucleotide preference as full-length Atx1 (de Chiara et al. 2003, Yue et al. 2001). Although binding seems to be sequence-specific, RNA targets of Atx1 AXH and the implications for the function of the native protein still remain to be investigated.

Finally, the AXH domain accounts for dimerization of the full-length protein (Burright et al. 1997, de Chiara et al. 2003). In addition, in vitro and in cell studies have shown that the isolated Atx1 AXH is able to aggregate spontaneously and favour the aggregation of the expanded full-length protein (de Chiara et al. 2005a, b). These findings indicate that the domain also plays a role in the SCA1 pathogenesis. The intrinsic ability of polyQ flanking domains to aggregate has been assessed also in recent studies on ataxin-3 and huntingtin exon-1 suggesting that the protein context modulates the disease and accounts for differences in the severity and age of onset given the same length of polyQ tract (Saunders and Bottomley 2009).

3.4 Towards the Determination of the Atx1 Interactome

More recently, using a stringent yeast two-hybrid screen, an interaction network was developed for 54 proteins involved in 23 inherited ataxias among which Atx1 (Lim et al. 2006). More than ~700 novel protein–protein interactions were identified. Some of them were verified in mammalian cells. It was found that many ataxia-causing proteins share interacting partners, a subset of which has been found to modify neurodegeneration in animal models. This interactome thus provides a tool for understanding pathogenic mechanisms common for this class of neurodegenerative disorders and for identifying candidate genes for inherited ataxias.

4 Proteasome Functions in SCA1

Another different but pathology-related aspect has attracted large attention in Atx1 studies. Alteration of the proteasome function in degrading misfolded proteins has been widely investigated for different proteins involved in neurodegenerative diseases. Several studies of how expanded and misfolded Atx1 enters in this pathway have thus been carried out.

In affected neurons of SCA1 patients and transgenic mice, mutant Atx1 accumulates in a single, ubiquitin-positive nuclear inclusion (Cummings et al. 1998). It was shown that these inclusions stain positively for the 20S proteasome and the DnaJ-like molecular chaperone HDJ-2/HSDJ and that overexpression of wild-type HDJ-2/HSDJ in HeLa cells decreases the frequency of Atx1 aggregation. These results suggest that protein misfolding is responsible for the nuclear aggregates seen in SCA1, and that overexpression of a DnaJ chaperone promotes recognition of a misfolded polyglutamine repeat protein, allowing its refolding and/or ubiquitin-dependent degradation.

Wild-type and mutant Atx1 with an expanded polyQ tract were shown to be polyubiquitinated equally well in transfected HeLa cells in vitro (Cummings et al. 1999), but the mutant form was three times more resistant to degradation suggesting that inhibition of proteasomal degradation promotes aggregation of mutant Atx1. Mouse Purkinje cells expressing mutant Atx1 but not the ubiquitin-protein ligase Ube3a showed fewer ubiquitin-positive nuclear inclusions than SCA1 mice, but the Purkinje cell pathology was markedly worse. The authors concluded that nuclear inclusions are not necessary to induce neurodegeneration, although impaired proteasomal degradation of mutant Atx1 may contribute to SCA1 pathogenesis.

Davidson et al. (2000) identified by the yeast two-hybrid system an ubiquitin-like nuclear protein A1Up as an Atx1-interacting protein (Davidson et al. 2000). Whereas yeast two-hybrid liquid culture assays indicated that A1Up interacts to a similar extent with wild-type and mutant Atx1, in the nucleus, A1Up co-localized with mutant Atx1. The presence of an N-terminal ubiquitin-like (Ubl) domain suggested that A1Up interaction with Atx1 might have a role in targeting the protein to the proteasome. In a further attempt to characterize A1Up and the significance of its interaction with Atx1, the authors showed that, similarly to other Ubl-containing proteins, the Ubl of A1Up is essential for the interaction of A1Up with the S5a subunit of the 19S proteasome and the interaction with the 19S proteasome was disrupted in the presence of Atx1 (Riley et al. 2004).

Also using a yeast two-hybrid system, Hong et al. found that the ubiquitin-specific protease USP7 binds to Atx1 and that the Atx1 C-terminus is essential for interaction (Hong et al. 2002). Liquid β-galactosidase assay and co-immunoprecipitation experiments revealed that the strength of the interaction between USP7 and Atx1 is influenced by the length of the polyQ tract; weaker interaction was observed in mutant Atx1 with longer polyQ tract and USP7 was not recruited to the mutant Atx1 aggregates in the Purkinje cells of SCA1 transgenic mice.

Another study based on the use of d2EGFP, a short-lived enhanced green fluorescent protein, investigated whether polyQ-expanded Atx1 affects the function of the proteasome (Park et al. 2005). It was shown that the expanded protein decreases the activity of the proteasome, implying that a disturbance in the ubiquitin-proteasome pathway is directly involved in the development of SCA1.

Finally, the role of CHIP, a chaperone Hsc-70 interacting protein that links the protein folding machinery with the ubiquitin-proteasome system, in protecting from Atx1-induced neurodegeneration has been investigated (Al-Ramahi et al. 2006). The authors found that CHIP interacts directly with both expanded and non-expanded Atx1 and co-localizes with Atx1 in nuclear inclusions. CHIP also promotes ubiquitination of expanded Atx1 both in vitro and in cell culture. Over-expression of CHIP in a *Drosophila* model of SCA1 decreases the protein steady-state levels of both expanded and unexpanded Atx1 and suppresses their toxicity. CHIP plays a protective role in SCA1 neuropathology by targeting expanded Atx1 for proteasomal degradation in a chaperone dependent manner. These results were confirmed by an independent study which showed that CHIP associates also with normal Atx1 (Choi et al. 2007). By enhancing Atx1 ubiquitination, CHIP overexpression leads to a reduction of solubility which increases aggregate formation and accumulation of the protein in the insoluble fraction.

5 Conclusions

It is undoubted that our knowledge on the function(s) of Atx1 has steadily increased over the last decade. Yet, much remains to be learned before this knowledge can be translated into a reliable treatment of SCA1, certainly the most important target for our current studies.

Acknowledgments We would like to thank Rajesh P. Menon and Paola Giunti for helpful comments.

References

Al-Ramahi, I., Lam, Y.C., Chen, H.K., de Gouyon, B., Zhang, M., Perez, A.M., Branco, J., de Haro, M., Patterson, C., Zoghbi, H.Y. et al. (2006) CHIP protects from the neurotoxicity of expanded and wild-type ataxin-1 and promotes their ubiquitination and degradation. *J. Biol. Chem.* **281**: 26714–26724.

Banfi, S., Servadio, A., Chung, M.Y., Kwiatkowski, T.J., Jr., McCall, A.E., Duvick, L.A., Shen, Y., Roth, E.J., Orr, H.T. and Zoghbi, H.Y. (1994) Identification and characterization of the gene causing type 1 spinocerebellar ataxia. *Nat. Genet.* **7**: 513–520.

Bauer, P.O. and Nukina, N. (2009) The pathogenic mechanisms of polyglutamine diseases and current therapeutic strategies. *J. Neurochem.* **110**: 1737–1765.

Bolger, T.A., Zhao, X., Cohen, T.J., Tsai, C.C. and Yao, T.P. (2007) The neurodegenerative disease protein ataxin-1 antagonizes the neuronal survival function of myocyte enhancer factor-2. *J. Biol. Chem.* **282**: 29186–29192.

Bowman, A.B., Lam, Y.C., Jafar-Nejad, P., Chen, H.K., Richman, R., Samaco, R.C., Fryer, J.D., Kahle, J.J., Orr, H.T. and Zoghbi, H.Y. (2007) Duplication of Atxn1l suppresses SCA1 neuropathology by decreasing incorporation of polyglutamine-expanded ataxin-1 into native complexes. *Nat. Genet.* **39**: 373–379.

Burright, E.N., Davidson, J.D., Duvick, L.A., Koshy, B., Zoghbi, H.Y. and Orr, H.T. (1997) Identification of a self-association region within the SCA1 gene product, ataxin-1. *Hum. Mol. Genet.* **6**: 513–518.

Chen, H.K., Fernandez-Funez, P., Acevedo, S.F., Lam, Y.C., Kaytor, M.D., Fernandez, M.H., Aitken, A., Skoulakis, E.M., Orr, H.T., Botas, J. et al. (2003) Interaction of Akt-phosphorylated ataxin-1 with 14-3-3 mediates neurodegeneration in spinocerebellar ataxia type 1. *Cell* **113**: 457–468.

Chen, Y.W., Allen, M.D., Veprintsev, D.B., Lowe, J. and Bycroft, M. (2004) The structure of the AXH domain of spinocerebellar ataxin-1. *J. Biol. Chem.* **279**: 3758–3765.

Choi, J.Y., Ryu, J.H., Kim, H.S., Park, S.G., Bae, K.H., Kang, S., Myung, P.K., Cho, S., Park, B.C. and Lee do, H. (2007) Co-chaperone CHIP promotes aggregation of ataxin-1. *Mol. Cell Neurosci.* **34**: 69–79.

Cummings, C.J., Mancini, M.A., Antalffy, B., DeFranco, D.B., Orr, H.T. and Zoghbi, H.Y. (1998) Chaperone suppression of aggregation and altered subcellular proteasome localization imply protein misfolding in SCA1. *Nat. Genet.* **19**: 148–154.

Cummings, C.J., Reinstein, E., Sun, Y., Antalffy, B., Jiang, Y., Ciechanover, A., Orr, H.T., Beaudet, A.L. and Zoghbi, H.Y. (1999) Mutation of the E6-AP ubiquitin ligase reduces nuclear inclusion frequency while accelerating polyglutamine-induced pathology in SCA1 mice. *Neuron* **24**: 879–892.

Cvetanovic, M., Rooney, R.J., Garcia, J.J., Toporovskaya, N., Zoghbi, H.Y. and Opal, P. (2007) The role of LANP and ataxin 1 in E4F-mediated transcriptional repression. *EMBO Rep.* **8**: 671–677.

Davidson, J.D., Riley, B., Burright, E.N., Duvick, L.A., Zoghbi, H.Y. and Orr, H.T. (2000) Identification and characterization of an ataxin-1-interacting protein: A1Up, a ubiquitin-like nuclear protein. *Hum. Mol. Genet.* **9**: 2305–2312.

de Chiara, C., Giannini, C., Adinolfi, S., de Boer, J., Guida, S., Ramos, A., Jodice, C., Kioussis, D. and Pastore, A. (2003) The AXH module: an independently folded domain common to ataxin-1 and HBP1. *FEBS Lett.* **551**: 107–112.

de Chiara, C., Menon, R.P., Adinolfi, S., de Boer, J., Ktistaki, E., Kelly, G., Calder, L., Kioussis, D. and Pastore, A. (2005a) The AXH domain adopts alternative folds the solution structure of HBP1 AXH. *Structure* **13**: 743–753.

de Chiara, C., Menon, R.P., Dal Piaz, F., Calder, L. and Pastore, A. (2005b) Polyglutamine is not all: the functional role of the AXH domain in the ataxin-1 protein. *J. Mol. Biol.* **354**: 883–893.

de Chiara, C., Menon, R.P., Strom, M., Gibson, T.J. and Pastore, A. (2009) Phosphorylation of s776 and 14-3-3 binding modulate ataxin-1 interaction with splicing factors. *PLoS One* **4**: e8372.

Emamian, E.S., Kaytor, M.D., Duvick, L.A., Zu, T., Tousey, S.K., Zoghbi, H.Y., Clark, H.B. and Orr, H.T. (2003) Serine 776 of ataxin-1 is critical for polyglutamine-induced disease in SCA1 transgenic mice. *Neuron* **38**: 375–387.

Fernandez-Funez, P., Nino-Rosales, M.L., de Gouyon, B., She, W.C., Luchak, J.M., Martinez, P., Turiegano, E., Benito, J., Capovilla, M., Skinner, P.J. et al. (2000) Identification of genes that modify ataxin-1-induced neurodegeneration. *Nature* **408**: 101–106.

Genis, D., Matilla, T., Volpini, V., Rosell, J., Davalos, A., Ferrer, I., Molins, A. and Estivill, X. (1995) Clinical, neuropathologic, and genetic studies of a large spinocerebellar ataxia type 1 (SCA1) kindred: (CAG)n expansion and early premonitory signs and symptoms. *Neurology* **45**: 24–30.

Goold, R., Hubank, M., Hunt, A., Holton, J., Menon, R.P., Revesz, T., Pandolfo, M. and Matilla-Duenas, A. (2007) Down-regulation of the dopamine receptor D2 in mice lacking ataxin 1. *Hum. Mol. Genet.* **16**: 2122–2134.

Hong, S., Kim, S.J., Ka, S., Choi, I. and Kang, S. (2002) USP7, a ubiquitin-specific protease, interacts with ataxin-1, the SCA1 gene product. *Mol. Cell. Neurosci.* **20**: 298–306.

Irwin, S., Vandelft, M., Pinchev, D., Howell, J.L., Graczyk, J., Orr, H.T. and Truant, R. (2005) RNA association and nucleocytoplasmic shuttling by ataxin-1. *J. Cell Sci.* **118**: 233–242.

Jayaraman, M., Kodali, R. and Wetzel, R. (2009) The impact of ataxin-1-like histidine insertions on polyglutamine aggregation. *Protein Eng. Des. Sel.* **22**: 469–478.

Jodice, C., Malaspina, P., Persichetti, F., Novelletto, A., Spadaro, M., Giunti, P., Morocutti, C., Terrenato, L., Harding, A.E. and Frontali, M. (1994) Effect of trinucleotide repeat length and parental sex on phenotypic variation in spinocerebellar ataxia I. *Am. J. Hum. Genet.* **54**: 959–965.

Johnson, E.S. (2004) Protein modification by SUMO. *Annu. Rev. Biochem.* **73**: 355–382.

Jorgensen, N.D., Andresen, J.M., Lagalwar, S., Armstrong, B., Stevens, S., Byam, C.E., Duvick, L.A., Lai, S., Jafar-Nejad, P., Zoghbi, H.Y. et al. (2009) Phosphorylation of ATXN1 at Ser776 in the cerebellum. *J. Neurochem.* **110**: 675–686.

Klement, I.A., Skinner, P.J., Kaytor, M.D., Yi, H., Hersch, S.M., Clark, H.B., Zoghbi, H.Y. and Orr, H.T. (1998) Ataxin-1 nuclear localization and aggregation: role in polyglutamine-induced disease in SCA1 transgenic mice. *Cell* **95**: 41–53.

Klement, I.A., Zoghbi, H.Y. and Orr, H.T. (1999) Pathogenesis of polyglutamine-induced disease: a model for SCA1. *Mol. Genet. Metab.* **66**: 172–178.

Lam, Y.C., Bowman, A.B., Jafar-Nejad, P., Lim, J., Richman, R., Fryer, J.D., Hyun, E.D., Duvick, L.A., Orr, H.T., Botas, J. et al. (2006) ATAXIN-1 interacts with the repressor Capicua in its native complex to cause SCA1 neuropathology. *Cell* **127**: 1335–1347.

Lesage, F., Hugnot, J.P., Amri, E.Z., Grimaldi, P., Barhanin, J. and Lazdunski, M. (1994) Expression cloning in K+ transport defective yeast and distribution of HBP1, a new putative HMG transcriptional regulator. *Nucl. Acids Res.* **22**: 3685–3688.

Lim, J., Hao, T., Shaw, C., Patel, A.J., Szabo, G., Rual, J.F., Fisk, C.J., Li, N., Smolyar, A., Hill, D.E. et al. (2006) A protein-protein interaction network for human inherited ataxias and disorders of Purkinje cell degeneration. *Cell* **125**: 801–814.

Lim, J., Crespo-Barreto, J., Jafar-Nejad, P., Bowman, A.B., Richman, R., Hill, D.E., Orr, H.T. and Zoghbi, H.Y. (2008) Opposing effects of polyglutamine expansion on native protein complexes contribute to SCA1. *Nature* **452**: 713–718.

Lin, X., Antalffy, B., Kang, D., Orr, H.T. and Zoghbi, H.Y. (2000) Polyglutamine expansion down-regulates specific neuronal genes before pathologic changes in SCA1. *Nat. Neurosci.* **3**: 157–163.

Masino, L., Nicastro, G., Menon, R.P., Dal Piaz, F., Calder, L. and Pastore, A. (2004) Characterization of the structure and the amyloidogenic properties of the Josephin domain of the polyglutamine-containing protein ataxin-3. *J. Mol. Biol.* **344**: 1021–1035.

Matilla-Duenas, A., Goold, R. and Giunti, P. (2008) Clinical, genetic, molecular, and pathophysiological insights into spinocerebellar ataxia type 1. *Cerebellum* **7**: 106–114.

Matilla, A., Koshy, B.T., Cummings, C.J., Isobe, T., Orr, H.T. and Zoghbi, H.Y. (1997) The cerebellar leucine-rich acidic nuclear protein interacts with ataxin-1. *Nature* **389**: 974–978.

Mizutani, A., Wang, L., Rajan, H., Vig, P.J., Alaynick, W.A., Thaler, J.P. and Tsai, C.C. (2005) Boat, an AXH domain protein, suppresses the cytotoxicity of mutant ataxin-1. *EMBO J.* **24**: 3339–3351.

Murzin, A.G. (1993) OB(oligonucleotide/oligosaccharide binding)-fold: common structural and functional solution for non-homologous sequences. *EMBO J.* **12**: 861–867.

Mushegian, A.R., Bassett, D.E., Jr., Boguski, M.S., Bork, P. and Koonin, E.V. (1997) Positionally cloned human disease genes: patterns of evolutionary conservation and functional motifs. *Proc. Natl. Acad. Sci. USA* **94**: 5831–5836.

Okazawa, H., Rich, T., Chang, A., Lin, X., Waragai, M., Kajikawa, M., Enokido, Y., Komuro, A., Kato, S., Shibata, M. et al. (2002) Interaction between mutant ataxin-1 and PQBP-1 affects transcription and cell death. *Neuron* **34**: 701–713.

Orr, H.T., Chung, M.Y., Banfi, S., Kwiatkowski, T.J., Jr., Servadio, A., Beaudet, A.L., McCall, A.E., Duvick, L.A., Ranum, L.P. and Zoghbi, H.Y. (1993) Expansion of an unstable trinucleotide CAG repeat in spinocerebellar ataxia type 1. *Nat. Genet.* **4**: 221–226.

Orr, H.T. and Zoghbi, H.Y. (2001) SCA1 molecular genetics: a history of a 13 year collaboration against glutamines. *Hum. Mol. Genet.* **10**: 2307–2311.

Orr, H.T. and Zoghbi, H.Y. (2007) Trinucleotide repeat disorders. *Annu. Rev. Neurosci.* **30**: 575–621.

Park, Y., Hong, S., Kim, S.J. and Kang, S. (2005) Proteasome function is inhibited by polyglutamine-expanded ataxin-1, the SCA1 gene product. *Mol. Cells* **19**: 23–30.

Riley, B.E., Xu, Y., Zoghbi, H.Y. and Orr, H.T. (2004) The effects of the polyglutamine repeat protein ataxin-1 on the UbL-UBA protein A1Up. *J. Biol. Chem.* **279**: 42290–42301.

Riley, B.E., Zoghbi, H.Y. and Orr, H.T. (2005) SUMOylation of the polyglutamine repeat protein, ataxin-1, is dependent on a functional nuclear localization signal. *J. Biol. Chem.* **280**: 21942–21948.

Saunders, H.M. and Bottomley, S.P. (2009) Multi-domain misfolding: understanding the aggregation pathway of polyglutamine proteins. *Protein Eng. Des. Sel.* **22**: 447–451.

Seeler, J.S. and Dejean, A. (2003) Nuclear and unclear functions of SUMO. *Nat. Rev. Mol. Cell Biol.* **4**: 690–699.

Serra, H.G., Byam, C.E., Lande, J.D., Tousey, S.K., Zoghbi, H.Y. and Orr, H.T. (2004) Gene profiling links SCA1 pathophysiology to glutamate signaling in Purkinje cells of transgenic mice. *Hum. Mol. Genet.* **13**: 2535–2543.

Serra, H.G., Duvick, L., Zu, T., Carlson, K., Stevens, S., Jorgensen, N., Lysholm, A., Burright, E., Zoghbi, H.Y., Clark, H.B. et al. (2006) RORalpha-mediated Purkinje cell development determines disease severity in adult SCA1 mice. *Cell* **127**: 697–708.

Shao, J. and Diamond, M.I. (2007) Polyglutamine diseases: emerging concepts in pathogenesis and therapy. *Hum. Mol. Genet.* **16 Spec No. 2**: R115–123.

Tsai, C.C., Kao, H.Y., Mitzutani, A., Banayo, E., Rajan, H., McKeown, M. and Evans, R.M. (2004) Ataxin 1, a SCA1 neurodegenerative disorder protein, is functionally linked to the silencing mediator of retinoid and thyroid hormone receptors. *Proc. Natl. Acad. Sci. USA* **101**: 4047–4052.

Tsuda, H., Jafar-Nejad, H., Patel, A.J., Sun, Y., Chen, H.K., Rose, M.F., Venken, K.J., Botas, J., Orr, H.T., Bellen, H.J. et al. (2005) The AXH domain of Ataxin-1 mediates neurodegeneration through its interaction with Gfi-1/Senseless proteins. *Cell* **122**: 633–644.

Yue, S., Serra, H.G., Zoghbi, H.Y. and Orr, H.T. (2001) The spinocerebellar ataxia type 1 protein, ataxin-1, has RNA-binding activity that is inversely affected by the length of its polyglutamine tract. *Hum. Mol. Genet.* **10**: 25–30.

Zoghbi, H.Y. (2000) Spinocerebellar ataxias. *Neurobiol. Dis.* **7**: 523–527.

Zoghbi, H.Y. and Orr, H.T. (2000) Glutamine repeats and neurodegeneration. *Annu. Rev. Neurosci.* **23**: 217–247.

Zoghbi, H.Y. and Orr, H.T. (2009) Pathogenic mechanisms of a polyglutamine mediated neurodegenerative disease: SCA1. *J. Biol. Chem.* **284**: 7425–1749.

Phase Plate Electron Microscopy

Kuniaki Nagayama

Abstract An electron microscope enhanced with phase plates has practical advantages, particular for biological electron microscopy. It permits collecting high-contrast images of close-to-life biological structures with cryo-fixation and without harsh sample preparations including staining. Here we describe the state of the art of phase plate electron microscopy. Focuses are given to methodological aspects and biological applications with two dimensional and three-dimensional imaging.

Keywords Electron microscopy · Tomography · Phase plate · Phase contrast · Thin-film · Zernike phase contrast · Hilbert differential contrast · Cryo-fixation

1 Introduction

As transmitted waves (either light or electrons) pass through a sample under a microscope, the waves may be absorbed (changing their amplitude) and/or refracted (changing their phase). Samples that absorb transmitted waves to a high degree produce microscopic images primarily through alteration in wave amplitude. With transparent samples (called "phase objects" in TEM), amplitude contrast does not occur and contrast arises instead through phase alteration. In light microscopy, the first phase visualization technique appeared in the Schlieren optics at the end of the nineteenth century, followed by the appearance of Zernike phase contrast (ZPC) (Zernike 1942) and the Smith/Nomarski differential interference contrast (DIC) (Smith 1947, Nomarski 1952) methods. For an electron microscope, Sherzer developed the other way, defocus phase contrast (DPC) (Scherzer 1949) by taking an advantage of the big gap between the very short wave-length of electron waves utilized and the required spatial resolution. There have been many attempts to use phase plates that resemble those developed by Zernike in the context of TEM. All of

K. Nagayama (✉)
National Institute for Physiological Sciences, Okazaki City, Aichi 444-8787, Japan; The Graduate University for Advanced Studies, School of Physiological Sciences, Hayamacho, Kanagawa 240-0193, Japan
e-mail: nagayama@nips.ac.jp

J. Brnjas-Kraljević, G. Pifat-Mrzljak (eds.), *Supramolecular Structure and Function 10*, DOI 10.1007/978-94-007-0893-8_6,

these demonstrated some improvement in image contrast; however, various difficulties associated with manufacturing and charging of the phase plates prevented the techniques from being widely adopted. Recent research conducted by a Japanese group that devoted considerable time to phase plate development may solve most of the difficulties and make the use of phase plates practical (Nagayama 2005).

1.1 Defocus Phase Contrast TEM

Figure 1a demonstrates the process employed in conventional DPC-TEM (defocus phase contrast-transmission electron microscope). A circular aperture at the back focal plane (BFP) of the objective lens limits electrons contributing to image formation by intercepting those scattered outside the radius of the aperture. Two types of contrast generation are traditionally employed in the observation of biological samples by conventional TEM.

If the sample is stained using heavy elements, some of the incident electrons are scattered at high angles and thus intercepted by the objective aperture; this produces amplitude contrast in the final image (called as scattering contrast). This technique increases contrast but decreases resolution by introducing various artifacts in structural details.

Phase contrast can be introduced by defocusing the objective lens (refer to Fig. 1a). Depending on the degree of defocus, the spectral characteristics of phase information transfer change. Figure 1b shows a plot of the phase contrast transfer function (CTF). The CTF describes the amount of optical information transfer as a function of the spatial frequency. It oscillates with an initial value of zero at the origin of the frequency or the centre of the diffraction space (k-space), thus acting as a band-pass filter. Low frequency components are highly suppressed, leading to overall low contrast in images of unstained biological samples. Increasing the defocus value can improve the low-frequency transfer but greatly reduces the upper frequency bound and hence the resolution.

1.2 Zernike Phase Contrast TEM

Figure 1c outlines the design of a Zernike phase contrast TEM (ZPC-TEM). The only difference between ZPC-TEM and conventional TEM is the presence of a phase plate at the BFP, where the aperture supports the thin film of the plate. The thickness of the phase plate film is adjusted to $\pi/2$ (90°) phase shift. Mathematically, the BFP corresponds to a two-dimensional Fourier space characterizing the diffraction or the spatial frequency. Thus, the manipulation of frequency components at the BFP by the phase plates is equivalent to spatial filtering that is in turn able to manage phase contrast. The contrast transfer theory for ZPC-TEM, however, is beyond the scope of this paper and is developed elsewhere (Scherzer 1949, Nagayama 1999). The CTF describing the characteristics of ZPC-TEM is shown in Fig. 1d. Uniform

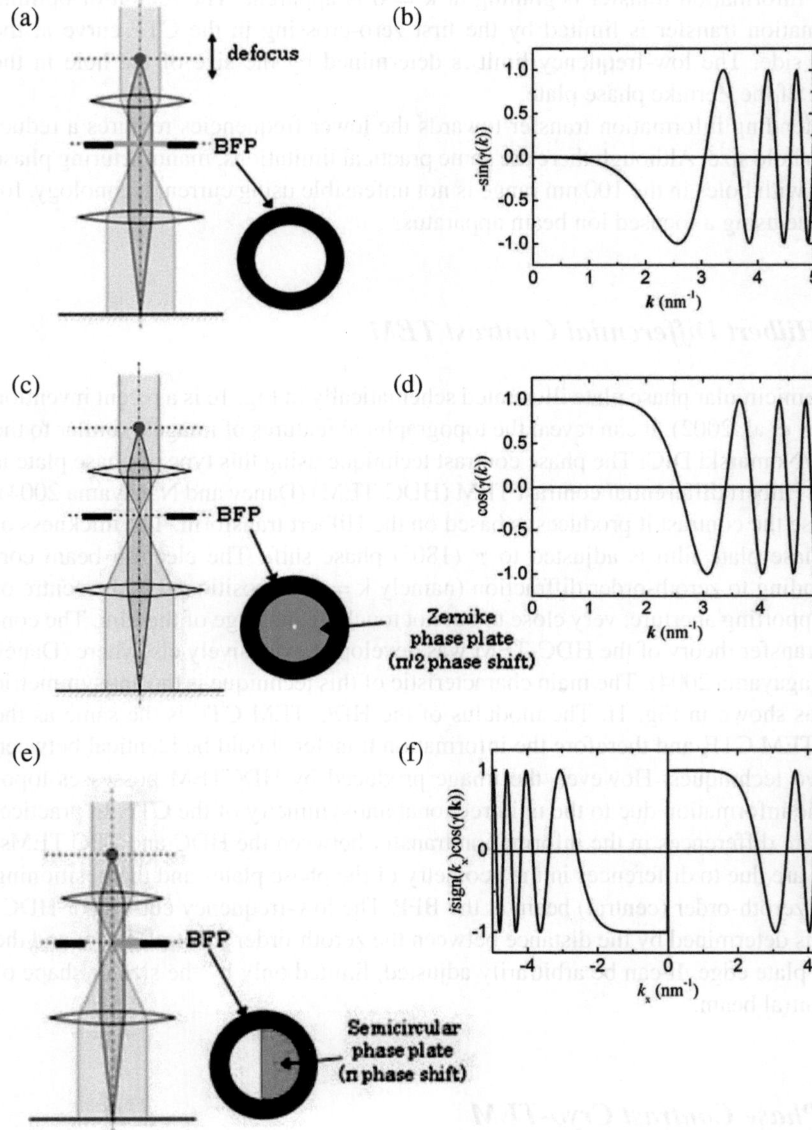

Fig. 1 Three types of phase contrast methods. (**a**) Schematic illustrating the defocus phase contrast (DPC) with conventional TEM, in which contrast is adjusted by altering the defocus. (**b**) In-focus contrast transfer function (CTF) corresponding to DPC plotted against the modulus of the spatial frequency (k). (**c**) Schematic illustrating ZPC using a Zernike phase plate set at the back focal plane (BFP). (**d**) In-focus CTF corresponding to ZPC plotted against the modulus of the spatial frequency (k) at the BFP. (**e**) Schematic illustrating the HDC using a Hilbert phase plate (semicircular phase plate) set at the BFP. (**f**) In-focus CTF corresponding to HDC plotted against a unidirectional spatial frequency (k_X). (**b, d, f**) 300 kV, $\lambda = 0.001968$ nm and Cs = 3 mm (adapted from fig. 1 of Danev and Nagayama 2006)

intact information transfer beginning at $k = 0$ is apparent. The region of optimal information transfer is limited by the first zero-crossing in the CTF curve at the upper side. The low-frequency limit is determined by the size of the hole in the centre of the Zernike phase plate.

Extending information transfer towards the lower frequencies requires a reduction in hole size. Although there are some practical limitations, manufacturing phase plates with holes in the 100 nm range is not unfeasible using current technology, for instance using a focused ion beam apparatus.

1.3 Hilbert Differential Contrast TEM

The semicircular phase plate illustrated schematically in Fig. 1e is a recent invention (Danev et al. 2002). It can reveal the topographical features of images, similar to the Smith/Nomarski DIC. The phase contrast technique using this type of phase plate is called Hilbert differential contrast TEM (HDC-TEM) (Danev and Nagayama 2004), because the contrast it produces is based on the Hilbert transform. The thickness of the phase plate film is adjusted to π (180°) phase shift. The electron beam corresponding to zeroth-order diffraction (namely $k = 0$) is positioned at the centre of the supporting aperture, very close to but not touching the edge of the film. The contrast transfer theory of the HDC-TEM was developed extensively elsewhere (Danev and Nagayama 2004). The main characteristic of this technique is the antisymmetric CTF as shown in Fig. 1f. The modulus of the HDC-TEM CTF is the same as the ZPC-TEM CTF, and therefore the information transfer should be identical between the two techniques. However, the image produced by HDCTEM possesses topographic information due to the unidirectional antisymmetry of the CTF. In practice, there are differences in the information transfer between the HDC and ZPC TEMs. These are due to differences in the geometry of the phase plates and the positioning of the zeroth-order (central) beam at the BFP. The low-frequency cut-off for HDC-TEM is determined by the distance between the zeroth-order (central) beam and the phase plate edge. It can be arbitrarily adjusted, limited only by the size or shape of the central beam.

1.4 Phase Contrast Cryo-TEM

Sample preparation has always been one of the most challenging and crucial issues in applying TEM to biological samples. Three properties of biological specimens ((i) they are made of aqueous media, which are inappropriate for vacuum conditions, (ii) they consist of light elements (H, C, N and O), which weakly diffract electron waves, and (iii) their internal structure is frequently very complicated) make their sample preparation difficult. A significant amount of time and effort has been devoted to resolving these issues, and a standard method has now been established, as shown in Fig. 2a–c. This method has been widely used, but it does have several

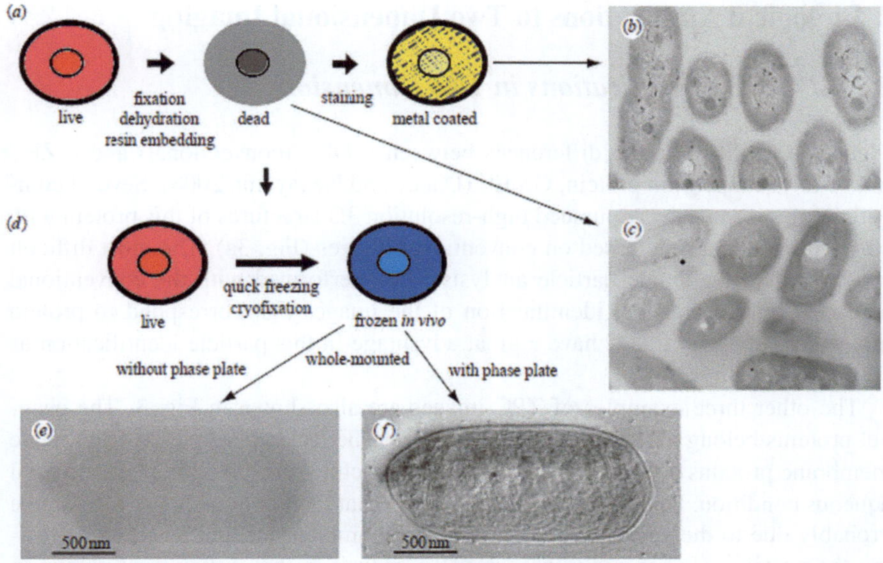

Fig. 2 (**a**) A conventional sample preparation and (**d**) an improved sample preparation using a cryotechnique. TEM micrographs associated with the preparation techniques are compared. (**b**) A conventional DPC-TEM image of a cyanobacterial cell sample prepared with conventional sample treatment with staining (100 kV). (**c**) A conventional DPC-TEM images of a cyanobacterial cell sample prepared with conventional sample treatment without staining (100 kV). (**d**) A sample preparation using quick freezing. (**e**) A conventional DPC-TEM image for a vitrified cyanobacterial cell with deep defocusing (300 kV). (**f**) A HDC-TEM image for a vitrified cyanobacterial cell with a Hilbert phase plate (300 kV) (images [**e, f**] were taken from fig. 1 of Kaneko et al. 2005)

drawbacks, which include its destructiveness to the sample, possible introduction of artifacts and the significant amount of time required.

The problems inherent in the traditional approach to sample preparation have largely been solved by introducing cryotechniques. Rapidly frozen vitrified (ice-embedded) specimens are much more probable than traditionally prepared samples to reflect the intrinsicinternal structures of biological samples without interference from artifacts.

The cryotechnique can be even more valuable when combined with a novel contrast enhancement technique using phase plates, and as such may give a useful artifact-free method to biological electron microscopists. Figure 2d–f provide an example depicting the cryopreparation process and the resultant images taken with DPC-TEM and HDC-TEM, respectively. Given that freezing is less harmful than other methods such as dehydration, this method is much more likely to be depicting the cyanobacterial cell in its in vivo state than those involving other preservation techniques. The fact that HDC-TEM also provides images with high contrast allows the intact fine structures to be easily recognized, as shown in Fig. 2f.

2 Biological Applications to Two Dimensional Imaging

2.1 ZPC-TEM Applications in Two Dimensions

Figure 3a, b illustrate the differences between a DPC (conventional) and a ZPC image for a chaperonin protein, GroEL (Danev and Nagayama 2008). Several competing groups intensively pursued high-resolution 3D structures of this protein with single particle analysis based on conventional images (Fig. 3a). The most difficult component of the single particle analysis when performed with the conventional approach is the first step, identification of the images that correspond to protein molecules. Images of ZPC have a great advantage in this particle identification as shown in Fig. 3b.

The other three examples of ZPC images are also shown in Fig. 3. The channel proteins belong to the most difficult group to be tackled with TEM as they are membrane proteins against solubilization and structural preservation in an artificial aqueous condition. Particularly it is well-known that their contrast in TEM is vague probably due to the smeared boundary between ambient ice and detergents dressing the protein. ZPC seems to be one of an answer to this difficulty as shown in Fig. 3d, d' (Shigematsu et al. 2010). The visibility of such a channel protein as having relatively small molecular weight, rat TRPV4, is remarkable.

When much larger biological specimens are treated, the advantage using ZPC is not decisive. Nevertheless as seen in the examples with a virus, T4 phage, there is a concrete gain in the structural elucidations as shown in Fig. 3e, f (Danev and Nagayama 2010). Keep in mind that the images in Fig. 3e, f are shown using the same intensity scale. The higher contrast of the Zernike image is due to improved transfer for the low spatial frequencies as mentioned. In addition, fine fiber-like structures protruding from the T4 phage surface are clearly recognized (Fig. 3f) because of the uniform transfer for a wide portion of the spatial frequency spectrum. Insets in Fig. 3e, f show the amplitudes of the Fourier transforms of the images. The DPC spectrum (inset in Fig. 3e) shows moderate presence of low frequency information around the center of the spectrum and ring-shaped areas of reduced amplitude due to CTF zeros. The ZPC-TEM spectrum (inset in Fig. 3f) does not exhibit CTF zeros and has very strong presence of low frequency information.

The last example to illustrate the comparison of DPC and ZPC is a quickly frozen whole primary cultured neuronal cell (unsectioned) derived from the cerebral cortices and hippocampi of E16 mice (Fig. 3g, h) (Fukuda et al. 2009). Fine details such as cytoskeletal filaments inside the cell are much more accessible with ZPC. Particularly, vertically running fine fibers recognized inside the cell (Fig. 3h) have been interpreted as intermediate filaments by their width.

Fig. 3 Comparison of DPC and ZPC images for various ice-embedded samples. (**a**) A DPC image (300 kV) for GroEL. The *inset* is a diffractogram obtained by Fourier-transform of the image. A typical sine CTF (refer to Fig. 1b) is shown (taken from fig. 2 of Danev and Nagayama 2008). (**b**) A ZPC image (300 kV) for GroEL. The *inset* is a diffractogram obtained by Fourier-transform of the image. A typical cosine CTF (refer to Fig. 1d) is shown (taken from fig. 2 of Danev and Nagayama

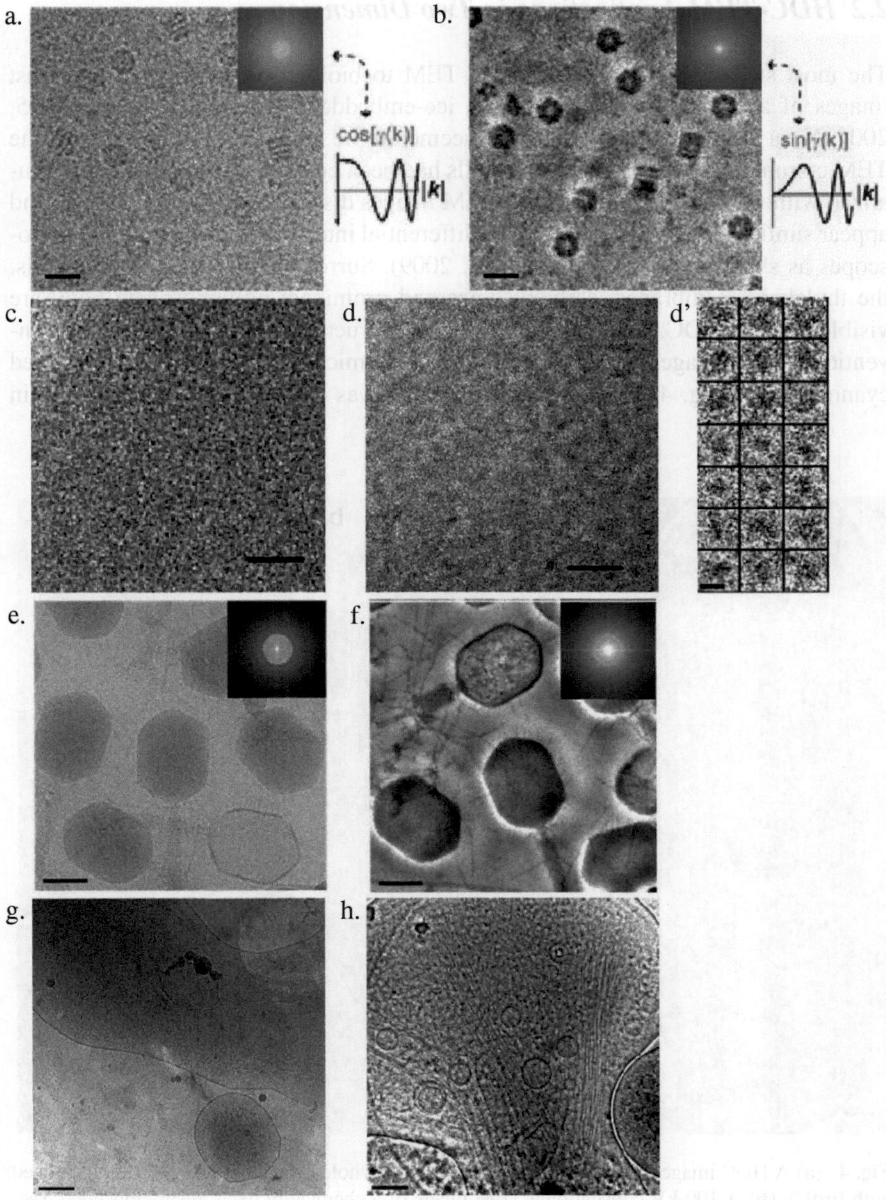

Fig. 3 (continued) 2008). *Bars* in **a** and **b** are 20 nm. (**c**) A DPC image (300 kV) for a calcium channel protein, rat TRPV4. *Bar* = 50 nm. (**d**) A ZPC image (300 kV) of rat TRPV4. *Bars* in **c** and **d** are 50 nm (taken from fig. 3 of Shigematsu et al. 2010). (**d'**) TRPV4 particles picked up from the image shown in **d**. *Bar* = 10 nm. (**e**) A DPC image (200 kV) for T4 phage (taken from fig. 2 of Danev and Nagayama 2010). (**f**) A ZPC image (200 kV) for T4 phage (taken from fig. 2 of Danev and Nagayama 2010). *Bars* in **e** and **f** are 50 nm. (**g**) A DPC image (300 kV) for a primary cultured neuronal cell (taken from fig. 5 of Fukuda et al. 2009). (**h**) A ZPC image (300 kV) for a primary cultured neuronal cell (taken from fig. 5 of Fukuda et al. 2009). *Bars* in **g** and **h** are 200 nm

2.2 HDC-TEM Applications in Two Dimensions

The most striking application of HDC-TEM to biology produced phase-contrast images of a whole-mounted cell in an ice-embedded state (Kaneko et al. 2005, 2007, Nitta et al. 2009). These results seemed to be completely unexpected in the TEM community, as whole-mounted cells had been considered too thick to be visualized without sectioning. The HDC-TEM images display topographic features and appear similar to images obtained with differential interference contrast light microscopes as shown in Fig. 4a (Nitta et al. 2009). Surrounded by smooth cell walls, the thylakoid membranes, carboxysomes and prominent polyphosphate body are visible through HDC. The identification of the structures was confirmed with conventional TEM images of ultrathin sections of chemically fixed and resin-embedded cyanobacteria (Fig. 4b). A further experiment was designed to identify DNA in

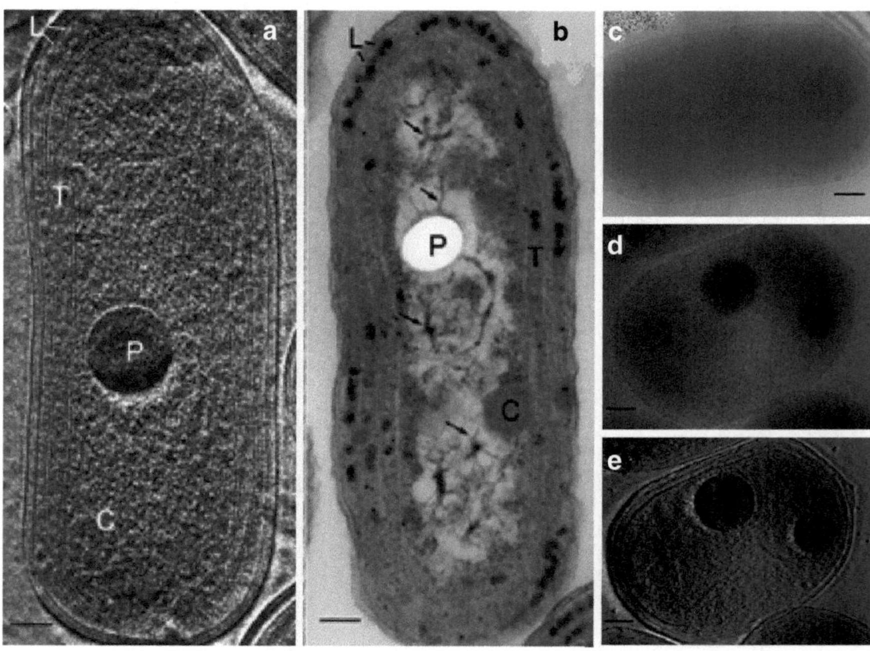

Fig. 4 (**a**) A HDC image (300 kV) for an ice-embedded whole cyanobacterial cell (not incubated with BrdU). (**b**) A 100-kV conventional TEM image of a chemically fixed, resin-embedded, sectioned cell (not incubated with BrdU). The sections were stained with uranyl acetate and lead citrate, which conferred electron density to strands assumed to be DNA. *C*, carboxysomes; *L*, lipid droplets; *P*, polyphosphate body (in **a**) or its residual hole (in **b**); *T*, thylakoid membranes; *arrows*: DNA fibres. (**c**) A HDC image (300 kV) image for an ice-embedded whole cyanobacterial cell incubated with BrdU for 2 h. (**d**) A DPC image (300 kV) for an ice-embedded whole cyanobacterial cell incubated with BrdU for 24 h. (**e**) A HDC image (300 kV) for the same ice-embedded whole cyanobacterial cell incubated with BrdU for 24 h. *Bars* = 100 nm (taken from figs. 1 and 4 of Nitta et al. 2009)

HDC-TEM images since DNA exhibits characteristic shapes in conventional TEM (Fig. 4b, arrows).

After rapidly growing cells were cultured in BrdU-containing media for 24 h, incorporation of BrdU into DNA was confirmed by fluorescent microscopy using FITC-labeled anti-BrdU antibodies and by EDX line analysis during TEM observation of resin embedded sections (Nitta et al. 2009). When BrdU incorporating whole cells were observed with TEM after rapid freezing, electron dense areas appeared in certain portions of the cells imaged at high under-focus without phase plate (Fig. 4d). No electron dense area except for the polyphosphate body, was obvious in cells incubated for only 2 h. When the HDC phase plate is applied, visualization of detailed ultrastructure of the electron dense area is greatly enhanced by its effect (Fig. 4e).

3 Biological Applications to Three Dimensional Imaging

3.1 ZPC-TEM Applications in Three Dimensions

The particle bigger, the image contrast from ambient ice is more prominent. This is the major cause why bigger protein particles are preferred in the single particle analysis. One of such an example is shown in Fig. 5a, b, where a tomographic reconstruction of a part of flagellar motor hook basal body (HBB) is shown based on DPC and ZPC cryo-tomography respectively (Hosogi et al. 2011). The contrast difference between sliced data of HBB with two schemes is clearly recognized, leading to an easy interpretation of internal structures in the complex protein.

The tomographically sliced image taken from DPC and ZPC tomograms of a T4 phage sample shown in Fig. 3c, d demonstrate the same higher contrast characteristics in ZPC as seen in Fig. 3e, f (Danev et al. 2010).

HDC tomography has also been recently reported (Barton et al. 2008).

4 Future Prospects and Conclusions

4.1 Electrostatic Charging

TEM pioneers noted that contamination on the surface of phase plates by insulating materials was the source of the charge. The phase plate itself is not charged when made of a conducting material such as carbon. Efforts for hunting after actual origins of contaminants clarified that there were three major sources for charged contaminants: organic materials, metal oxides, and inorganic materials, which were unavoidably integrated into or onto phase plates during the fabrication procedure (Nagayama 2005). For the final answer to settle the dilemma of fabricating or not fabricating, they employed an old idea – electrostatic shielding. The charge-induced potential can be shielded by wrapping charges with conductive material such as

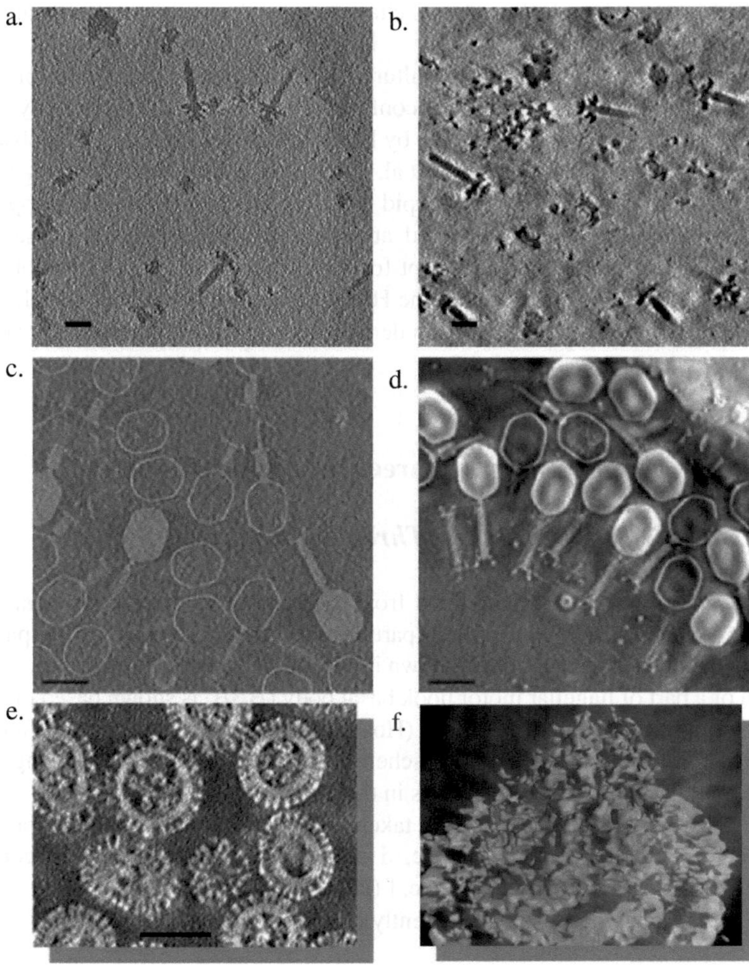

Fig. 5 (**a**) A DPC tomographically sliced image (200 kV) for a complex protein, hook basal body of flagellar motor (HBB) (adapted from fig. 1 of Hosogi et al. 2011). (**b**) A ZPC tomographically sliced image for a complex protein for HBB (adapted from fig. 1 of Hosogi et al. 2011). *Bars* in **a** and **b** are 50 nm. (**c**) A DPC tomographically sliced image (200 kV) for T4 phage (taken from fig. 4 of Danev and Nagayama 2010). (**d**) A ZPC tomographically sliced image (200 kV) for T4 phage (taken from fig. 4 of Danev and Nagayama 2010). *Bars* in **c** and **d** are 100 nm. (**e**) A ZPC tomographically sliced image to and influenza A virus sample. *Bar* = 100 nm. (**f**) A 3D structure of an influenza virus A graphically represented from one of virus tomograms shown in Fig. 5e (3D graphics prepared by Takaaki Takeda of National Institutes of Natural Sciences)

carbon. In the final step of phase-plate production, either sides of the phase plate, likely contaminated with organic materials, metal oxides, or inorganic materials, are coated with carbon in a vacuum evaporator. Consequently, once grounded, the electrostatic potential arising from charges is eliminated.

4.2 Next-Generation Phase Plates

One of the disadvantages in using thin-film phase-plates is electron loss due to electron scattering (Danev and Nagayama 2001, Nagayama 2005). In his pioneering work, Boersch proposed the other option for phase plates – electrostatic potential type (Boersch 1947). The sophisticated version of the Zernike phase plate containing a static ring electrode that can arbitrarily control the amount of phase shift was firstly proposed as the numerical work (Matumoto and Tonomura 1996) and recently many groups have reported that the electrostatic phase plate is technically tractable with the use of advanced microfabrication technology (Huang et al. 2006, Cambie et al. 2007, Majorovits et al. 2007, Shiue et al. 2009, Alloyeau et al. 2010) but efficient phase contrast has yet to be reported for biological samples in one part due to the large outer diameter of the central electrode ring blocking the lower frequency components and in the other part due to unavoidable charging of devices with complicated geometry. Phase plates are now continuously stimulating physicists as a novel kind of physical toys challenging various kinds of physical principles including Aharonov–Bohm effect (Aharonov and Bohm 1959, Nagayama 2008) and laser modulation approach (Danev and Nagayama 2010).

The trend in recent years has been observation of biological samples in their in vivo state, which usually means observation of vitrified samples. Phase contrast is the only way to extract information from such samples. However, maximum electron dose, which determines the strength of microscopic contrast, is limited when using unstained vitrified samples due to increased electron damage. Improvement of the phase contrast transfer of the electron microscope therefore becomes a burning issue. The recent developments in the application of thin-film phase plates to TEM show promising results as reported in this chapter. As more experiments are performed, more information gathered and more experience accumulated, these new techniques will gain further popularity and acknowledgement in the EM community.

Acknowledgments We owe the development and biological applications of phase contrast TEM with phase plates to the collaborators as follows. Development: Radostin Danev, Shozo Sugitani, Hiroshi Okawara, Toshiyuki Itoh, Toshikazu Honda, Toshiaki Suzuki, Yoshihiro Arai, Fumio Hosokawa, Sohei Motoki, Rasmus Schroeder and Michael Marko. Applications: Yasuko Kaneko, Koji Nitta, Hitoshi Nakamoto, Nobutaru Usuda, Kimie Atsuzawa, Ayami Nakazawa, Kiyokazu Kametani, Masashi Yamaguchi and Mitsutoshi Setou. This work was supported in part by a Grant-in-aid for Creative Scientific Research from MEXT, Japan and by Core Research for Evolutional Science and Technology (CREST) from JST, Japan.

References

Aharonov, Y. and Bohm, D. (1959) Significance of electromagnetic potentials in the quantum theory. *Phys. Rev.* **115**: 485–491.

Alloyeau, D., Hsieh, W.K., Anderson, E.H., Hilken, L., Benner, G., Meng, X., Chen, F.R. and Kisielowski, C. (2010) Imaging of soft and hard materials using a Boersh phase plate in a transmission electron microscope. *Ultramicroscopy* **110**: 563–570.

Barton, B., Joos, F. and Schröder, R.R. (2008) Improved specimen reconstruction by Hilbert phase contrast tomography. *J. Struct. Biol.* **164**: 210–220.

Boersch, H. (1947) Über die Kontraste von Atomen in Electronenmikroskop. *Z Naturforschg* **2a**: 615–633.

Cambie, R., Downing, K.H., Typke, D., Glaeser, R.M. and Jin, J. (2007) Design of a microfabricated, two-electrode phase-contrast element suitable for electron microscopy. *Ultramicroscopy* **107**: 329–339.

Danev, R. and Nagayama, K. (2001) Transmission electron microscopy with Zernike phase plate. *Ultramicroscopy* **88**: 243–252.

Danev, R., Okawara, H., Usuda, N., Kametani, K. and Nagayama, K. (2002) A novel phase-contrast transmission electron microscopy producing high-contrast topographic images of weak objects. *J. Biol. Phys.* **28**: 627–635.

Danev, R. and Nagayama, K. (2004) Complex observation in electron microscopy. IV. Reconstruction of complex object wave from conventional and half plane phase plate image pair. *J. Phys. Soc. Jpn.* **73**: 2718–2724.

Danev, R. and Nagayama, K. (2006) Applicability of thin film phase plates in biological electron microscopy. *Biophysics* **2**: 35–43.

Danev, R. and Nagayama, K. (2008) Single particle analysis based on Zernike phase contrast transmission electron microscopy. *J. Struct. Biol.* **161**: 211–218.

Danev, R. and Nagayama, K. (2010) Phase plates for transmission electron microscopy. *Meth. Enzymol.* **481**: 343–369.

Danev, R., Kanamaru, S., Marko, M. and Nagayama, K. (2010) Zernike phase contrast cryo-electron tomography. *J. Struct. Biol.* **171**(2): 174–181.

Fukuda, Y., Fukazawa, Y., Danev, R., Shigemoto, R. and Nagayama, K. (2009) Tuning of the Zernike phase plate for visualization of detailed ultrastructure in complex biological specimens. *J. Struct. Biol.* **168**: 476–484.

Hosogi, N., Shigematsu, H., Terashima, H., Honma, M. and Nagayama, K. (2011) Zernike phase contrast cryo-electron tomography of sodium-driven flagellar hook-basal bodies from *vibrio alginolytius. J. Struct. Biol.* **173**(1): 67–76.

Huang, S.H., Wang, W.J., Chang, C.S., Hwu, Y.K., Kai, J.J. and Chen, F.R. (2006) The fabrication and application of Zernike electrostatic phase plate. *J. Electron Mirosc.* **55**: 273–280.

Kaneko, Y., Danev, R., Nitta, K. and Nagayama, K. (2005) In vivo subcellular ultrastructures recognized with Hilbert-differential-contrast transmission electron microscopy. *J. Electron Microsc.* **54**: 79–84.

Kaneko, Y., Nitta, K. and Nagayama, K. (2007) Observation of in vivo DNA in ice embedded whole cyanobacterial cells by Hilbert differential contrast transmission electron microscopy (HDC-TEM). *Plasma Fusion Res.* **54**: 79–85.

Majorovits, E., Barton, B., Schultheiß, K., Perez-Willard, F., Gerthsen, D. and Schröder R.R. (2007) Optimizing phase contrast in transmission electron microscopy with an electrostatic (Boersch) phase plate. *Ultramicroscopy* **107**: 213–226.

Matumoto, T. and Tonomura, A. (1996) The phase constancy of electron waves travelling through Boersch's electrostatic phase plate. *Ultramicroscopy* **63**: 5–10.

Nagayama, K. (1999) Complex observation in electron microscopy. I. Basic scheme to surpass the Sherzer limit. *J. Phys. Soc. Jpn.* **68**: 811–822.

Nagayama, K. (2005) Phase contrast enhancement with phase plates in electron microscopy. *Ad. Imaging Electr. Phys.* **138**: 69–146.

Nagayama, K. (2008) Development of phase plates for electron microscopes and their biological application. *Eur. Biophys. J.* **37**: 345–358.

Nitta, K., Nagayama, K., Danev, R. and Kaneko, Y. (2009) Visualization of BrdU-labelled DNA in cyanobacterial cells by Hilbert differential contrast transmission electron microscopy. **234**: 118–123.

Nomarski, G. (1952) Interferométre á polarization. *French Patent* **1**: 059-123.

Scherzer, O. (1949) The theoretical resolution limit of the electron microscope. *J. Appl. Phys.* **20**: 20–29.

Shigematsu, H., Sokabe, T., Danev, R., Tominaga, M. and Nagayama, K. (2010) A 3.5-nm structure of rat TRPV4 cation channel revealed by Zernike phase-contrast cryo-EM. *J. Biol. Chem.* **285**: 11210–11218.

Shiue, J., Chang, C.S., Huang, S.H., Hsu, C.H., Tsai, J.S., Chang, W.H., Wu, Y.M., Lin, Y.C., Kuo, P.C., Huang, Y.S., Hwu, Y., Kai, J.J., T. F.G. and Chen, F.R. (2009) Phase TEM for biological imaging utilizing a Boersh electrostatic phase plate: theory and practice. *J. Electron Microsc.* **58**: 137–145.

Smith, F.H. (1947) Microscopes. British Patent 639 014, Class 97(i) CroupXX.

Zernike, F. (1942) Phase contrast, a new method for the microscopic observation of transparent objects. *Physica* **9**: 686–698, 974–986.

Deriving Biomedical Diagnostics from Spectroscopic Data

Ian C.P. Smith and Ray L. Somorjai

Abstract Biomedical spectroscopic experiments generate large volumes of data. For accurate, robust diagnostic tools the data must be analyzed for only a few characteristic observations per subject, and a large number of subjects must be studied. We describe here some of the current mathematical methods applied to this problem: Principal Component Analysis, Partial Least Squares, and the Statistical Classification Strategy. We demonstrate the application of these methods by three examples of their use in analyzing ^{1}H NMR spectra: screening for colon cancer, characterization of thyroid cancer, and distinguishing cancer from cholangitis in the biliary tract.

Keywords Biomedical spectroscopy · Multivariate methods · Classifiers · PC · PCA · SIMCA · Cancer

Abbreviations

FLD	Fisher's linear discriminant
FOBT	Fecal occult blood test
NMR	Nuclear magnetic resonance
PC	Principal component
PCA	Principal component analysis
PCR	Principal component regression
PLS	Partial least squares
PSC	Primary sclerosing cholangitis
SCS	Statistical classification strategy
SIMCA	Soft independent modelling of class analogies
WCVBST	Weighted cross validated bootstrap

I.C.P. Smith (✉)
Institute for Biodiagnostics, National Research Council Winnipeg, Winnipeg, MB, Canada R3B 1Y6
e-mail: ian.smith@nrc-cnrc.gc.ca

J. Brnjas-Kraljević, G. Pifat-Mrzljak (eds.), *Supramolecular Structure and Function 10*, DOI 10.1007/978-94-007-0893-8_7,

1 Introduction

Modern spectroscopic methods are being increasingly used to develop medical diagnostics. The methods lead to a plethora of data which defy analysis by conventional approaches. Over the past 10–20 years, a number of methods have been reported to simplify the data and analyze them accurately. The most commonly used approach, favored by many spectroscopists and chemometricians, is Soft Independent Modelling of Class Analogies (SIMCA), exemplified by the software SIMCA-P. SIMCA-P relies heavily on Principal Component Analysis (PCA), and for classification, on its supervised versions, Partial Least Squares (PLS) or Principal Component Regression (PCR). For short, we shall denote by "SIMCA" the entire corpus of the PCA/PLS/PCR-based methodology, detailed in reference Erikkson et al. (2001). The other approach is the Statistical Classification Strategy (SCS), developed at the Institute for Biodiagnostics (IBD). A detailed description of the SCS is found in the references Somorjai et al. (2004a, b). We shall introduce both approaches and compare their relative utility for the classification of nuclear magnetic resonance (NMR) spectra.

2 The Problems

Medical specimens or subjects are difficult to obtain in large numbers. Often, studies involving small numbers of subjects are reported, suggesting great potential for ultimate medical use, but regrettably, very few proceed further to actually demonstrate this and produce medically useful classifiers. This is the first problem.

Most biomedical spectra involve a large number of data points ("features"), many of which are not useful or even detrimental ("noise"), and only a few may actually be diagnostic. The second problem therefore is to extract the clinically meaningful features and limit their number (reduce the size of the feature space). A large number of features lead to overfitting and a non-robust classifier (one which fails on challenge by data not used to develop classifier).

Problem three involves identifying the appropriate mathematical method to develop the classifier. The conventional "scatter plots", used formerly in biomedical reports, are, at best based on univariate statistical tests (e.g., t-tests) and do little more than suggest possible potential for class separation. Actual confirmation of such potential requires appropriate multivariate methods and classifiers.

The top of Fig. 1 demonstrates the problem of insufficient number of samples. With very few samples, an apparently accurate separation is not difficult to achieve. Introduction of more samples will often produce a poorer separation because of the likelihood of filling the initially empty overlap between classes (bottom of Fig. 1). Finally, an adequate number of specimens may produce both a realistic, acceptable class separation and a potentially robust classifier. A rule of thumb is that the number of specimens must be roughly 10 times the number of discriminating features. Unfortunately, the converse is often reported.

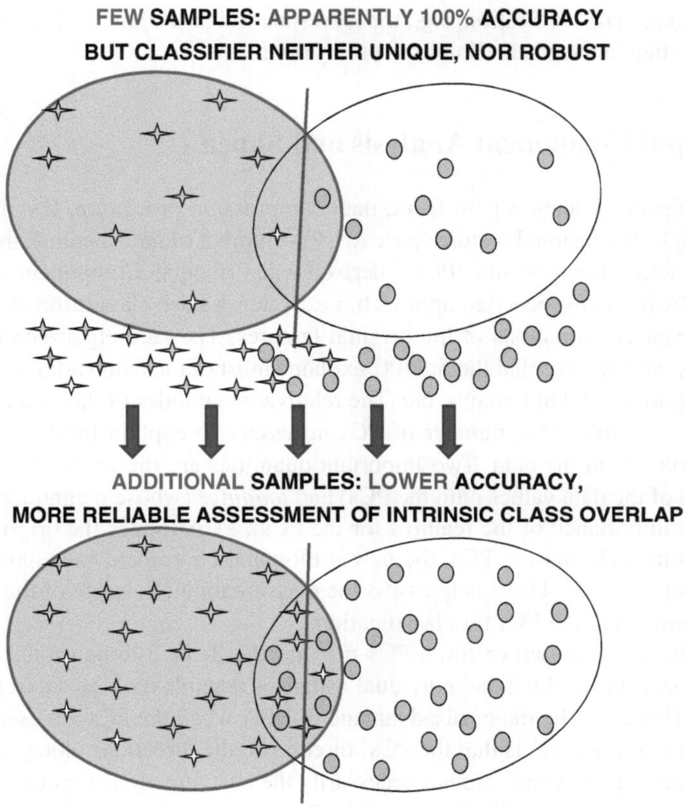

Fig. 1 Representation of the risks from reaching conclusion with a limited data set. Enlarging the number of subjects lowers the accuracy, but this is much closer to the true accuracy. The lower solution will also be more robust: challenging the resultant classifier with new specimens will yield accuracy equal to that found by an adequate classifier

Spectra must be prepared for analysis (preprocessing). This involves referencing to a standard frequency (chemical shift in NMR), scaling the features, and normalization to the area under the entire spectrum. Absorption or derivative spectra may be used. With absorption, the use of magnitude spectra simplifies analysis. Rank ordering of intensities has also proven useful, as did adding nonlinear terms to the feature space.

Determination of the best discriminating features in a spectrum is a critical requirement for classifier reliability and robustness. Ideally, the full dataset is first split into training and validation (monitoring) sets for classification. The optimal feature set is found using only the training set. The validation set serves as a control; it helps prevent excessive training and hence overfitting. If at all possible, an independent, blinded test set, one that didn't participate in the classifier development, should be challenged by the classifier to assess future performance realistically. When the dataset size is too small, even the training-test set split is unfeasible and some version of crossvalidation, e.g., the leave-one-out (LOO) method is commonly used, with attendant caveats (Somorjai et al. 2004a).

3 Principal Component Analysis and Simca

SIMCA's feature selection is in fact a data compression procedure. It reduces the original, high-dimensional feature space to a low number of new features, the uncorrelated principal components (PCs), derived via Principal Component Analysis (PCA). PCA is an unsupervised approach, i.e., it doesn't use class information. The PCs are linear combinations of the original features. They are eigenvectors of the data matrix, arranged so that the first PC explains most of the data variance, the second the majority of what remains, etc. The relative magnitudes of the data matrices' eigenvalues determine the number of PCs necessary to explain most of the variability (variance) in the data. Two important quantities are the *scores*, (orthogonal projections of the data values onto the PCs) and *loadings* (whose magnitudes reflect the relative importance of the features for the PCs). PCA rotates the original coordinate system such that the PCs, the new orthogonal coordinate axes, point along the maximal variance. These help assess the classification relevance of the original features comprising the PCs for classification.

Typically, the first two or three PCs provide the 2- or 3-dimensional reduced coordinate system in which the individual instances (samples) of the dataset may be displayed. However, the major disadvantage of PCA when the PCs are used as new features for classification is that the PCs' directions, the directions along which the dataset variance is maximal, are not necessarily the best directions for classification.

An important facet of SIMCA is that it allows computation of class-dependent PCAs. These are more flexible than the conventional PCs that don't use class-specific information. In particular, the relevance of different features and measures of class separation can be assessed more readily. The class-specific PCs are orthogonal within their class; they are not so between classes. This is not necessarily a disadvantage. However, in the SIMCA-based classification literature there doesn't seem to be any assessment of dependencies/correlations between PCs belonging to different classes. Such dependencies may be detrimental for robust, reliable classification.

4 Partial Least Squares (PLS) and Principal Component Regression (PCR)

PLS and PCR are two components of the SIMCA corpus that may be and are used for classification. (For two-class problems, linear classification and regression are equivalent.) They are more appropriate for classification than PCA, because the PLS/PCR PCs are determined by maximizing the covariance between the independent variables (the derived PCs) and the dependent ones (class membership). However, the caveats raised in connection with PCA still hold.

What is missing in SIMCA for classification? In principle, nothing. However, because the PCs are linearly "scrambled" versions of the original data points, direct identification and validation of the discriminating features ("biomarkers") is more difficult. Furthermore, for spectra the original features are single data points, hence cannot be readily and meaningfully interpreted as biomarkers. This lack of interpretability provided us the first impetus to develop the feature selection stage of the Statistical Classification Strategy (SCS). Furthermore, the general philosophy of the SIMCA approach, relying almost exclusively on PCA to reduce feature space dimensionality, independently of the data to be classified, led us to develop a data-driven classification strategy, without prior, preset assumptions. A flexible strategy is essential for biomedical spectroscopic data.

5 The Statistical Classification Strategy

This approach was developed in our institute. It consists of five stages. Of these, feature selection is critical. The particular method we developed and advocate was designed to produce features that retain spectral identity. In particular, for MR spectra the new features are averages of spectral intensities of adjacent data points of varying ranges (subregions), thus better representing specific peak locations and areas. A genetic algorithm-based optimal feature selector (Nikulin et al. 1998) is used; it optimizes both the number of discriminating subregions and their widths. The feature selection is "driven" by some preselected classifier that will eventually be employed for the actual classification (wrapper method). This approach leads to an optimal classifier-feature set pair. Because of its robustness, we frequently use Fisher's linear discriminant (FLD). FLD has the advantage of not only assigning samples to one of the classes, but also yielding a measure of the probability that they belong to that class. A more detailed mathematical description of the methods is given in reference (Nikulin et al. 1998).

When the sample size is small, in addition to or instead of the standard crossvalidation methods (e.g., leave-one-out), we generally use our bootstrapping-inspired approach, *weighted crossvalidated bootstrap* (WCVBST) (Somorjai 2009). The steps of this process are shown in Fig. 2. WCVBST was designed to create more robust classifiers, i.e., classifiers whose accuracy is more reliable for independent test sets. WCVBST selects randomly ~half of the samples to form a training set, develops with this an optimized classifier, and uses the remaining half of the samples to test the efficacy of this classifier. The process is repeated B times (typically, B = 5,000–10,000), always starting with the entire dataset. For all B random splits, the B sets of optimized classifier coefficients are saved. WCVBST is powerful because of the weighting we introduced: a *weighted average* of these B sets of coefficients produces the final, *single* classifier. The B weights are the ones found *not* for the training sets, but for the less optimistic *test sets*. We report classifier outcome as *class probability*.

When we are not satisfied with the accuracy of our classifier, we resort to classifier aggregation, a consensus technique (Kuncheva, 2004). One approach is to

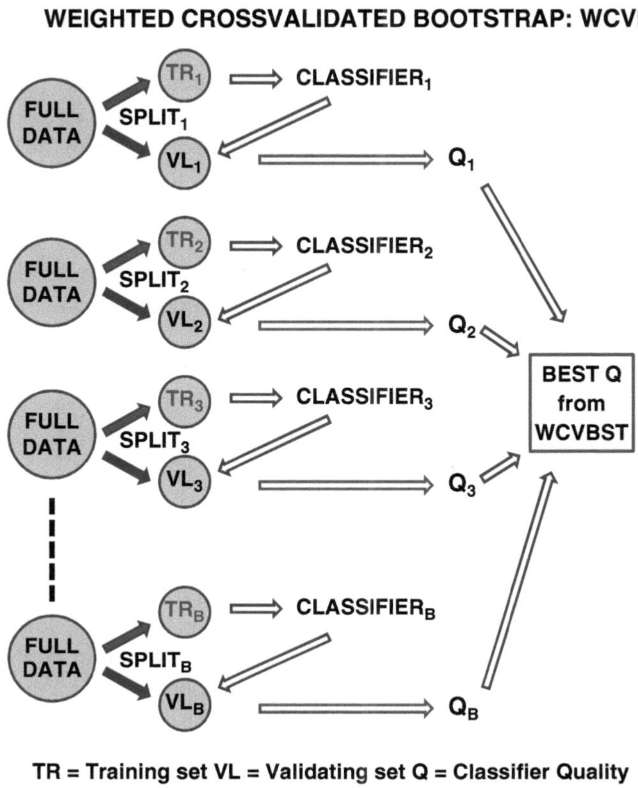

Fig. 2 Representation of the bootstrapping procedure. The data for the normals and the cancer patients are divided arbitrarily into two groups. A classifier is calculated from one, and tested by the other (TR_1 and VL_1). The entire data set is randomized and a second two sets composed (TR_2 and VL_2). This procedure continues many times (thousands) yielding many classifiers from which the weighted average, highest quality classifier is determined

use different representations of the spectra, such as absorption, first derivative, rank order, and develop full classifier-feature set pairs for each. Thus, we use a specific classifier, but on different types of feature sets. The classification outcomes are generally different and will emphasize different aspects of the spectra. Alternatively, we may select a specific feature set, perhaps based on prior information, but now develop, for this particular feature set, different classifiers, such as linear or quadratic discriminant analysis, some nearest neighbor classifier, neural nets, support vector machines, etc. In either case, we combine the classification outcomes to yield a consensus result (Somorjai et al. 2004a, b).

We emphasize that the SCS doesn't necessarily exclude PCs as possible features. In fact, PCs were also tried, with less success, in early versions of the SCS. However, unlike SIMCA, because of its data-driven philosophy and a built-in flexibility, the SCS doesn't confine itself to PCs as the only possible features. The strategy allows for and encourages experimentation with different combinations of preprocessing, feature selection, classifier choice and classifier aggregation. As an example, we generated class-specific feature sets (Somorjai et al. 2004b) that didn't require the use of class-dependent PCAs and PCs. Instead, in each class, lines that passed through pairs of samples were constructed and a sample point was assigned to the class whose line was closer to it. These lines are generalizations of PCs. They are more flexible, because neither orthogonality nor confinement to the direction of maximal variability was required and imposed.

6 Examples

6.1 Screening for Colon Cancer

In most countries colon cancer is the second worst killer among the cancers. If detected early, the prognosis is excellent (92%): detected late and 5-year survival drops to 6%. Clearly the way to prevent deaths is to find a method to test asymptomatic subjects and thus diagnosis early stages of colon cancer – subject screening. The current gold standard for screening is colonoscopy, where a flexible probe is inserted into the large intestine via the rectum and a visual assessment is made. The method is expensive, objectionable to many, and there is a risk of puncturing the intestinal wall and causing serious abdominal infection. A less expensive, less invasive method with a high accuracy is necessary. A start was made with the fecal occult blood test (FOBT), where a smear of stool is placed on a supportive strip that is subsequently tested chemically for the presence of hemoglobin. Its accuracy varies widely, from 40 to 80%, often due to the consumption of meat or red wine before the test. A risk of false positives leads to unnecessary colonoscopy, whereas a false negative leads to advancement of the stage of the colon cancer. Subject compliance, actually doing the test, is low.

In our institute, we have perfected a method to detect early colon cancer by performing NMR spectroscopy on fecal water, the supernate from a suspension of several grams of feces (Bezabeh et al. 2009). Typical ^{1}H spectra from unseparated suspensions are shown in Fig. 3. The spectra from the supernates show much greater resolution.

While differences between the spectra can be seen, it is necessary to determine the spectral regions most valuable for classification, step three of the SCS method. This is done by use of the genetic algorithm (see Section 5), which selects a few (e.g., 4–5) discriminating regions. The classification results are based on 523 individual cases (412 normal, 111 with colorectal cancer). An overall accuracy of 92% was obtained. Because we used FLD, we also obtained the probability that specimens belong in one of the classes. This probability may be viewed as the credibility ("crispness") of the class assignment. The high accuracy was obtained on samples having ≥75% crispness (405 specimens).

We are now in the process of moving this test into hospital use, which is expected to be complete in 2011. This study was performed in partnership with the MD Anderson Hospital, University of Texas, and the Health Sciences Centre, University of Manitoba.

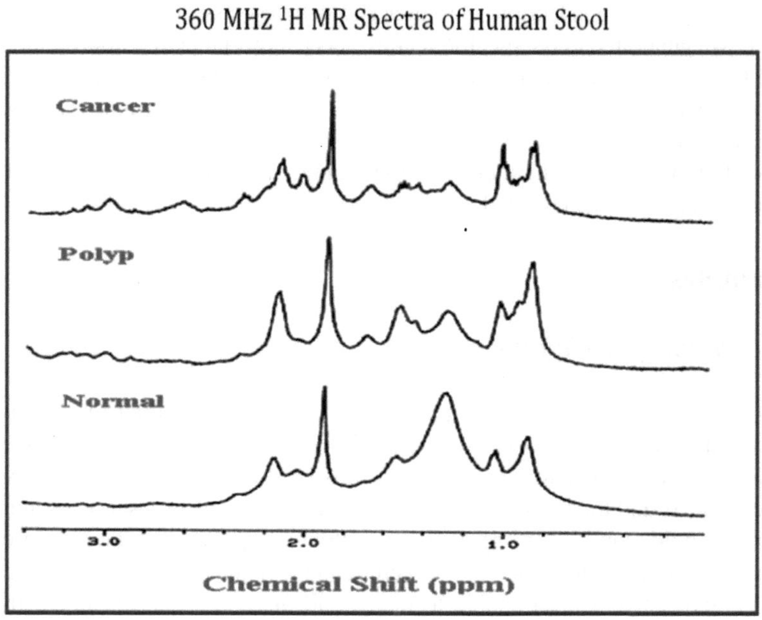

Fig. 3 NMR spectra (360 MHz, 27°C of fecal water from subjects and those with premalignant polyps and advanced colon cancer (Bezabeh et al. 2009)

6.2 Thyroid Cancer

Another example demonstrates the method of consensus analysis (Somorjai et al. 2004a, 1995). This involves analysis of the ^{1}H spectra of thyroid tissue biopsies by use of different classifier approaches, and then combining the resultant classifier outcomes to form a consensus result. These are the methods used in step five of the SCS (Somorjai et al. 2004a). The data were split into training and test sets. The three classifiers used are linear discriminant analysis, genetic programming, and neural networks. In this case we selected as the discriminating features the first ten PCs found by principal component analysis. Cross validation on the training sets was done by the standard leave-one-out method. For N samples, N slightly different classifiers are obtained; samples from the test set are assigned N times to one of the classes. The consensus result is the median of the N assignments. This early method is a precursor of the classifier aggregation step of the SCS; it produced an accuracy of 99% in a group of 107 thyroid biopsies from normal and various malignant cancers.

6.3 Biliary Cancer

In a recent publication (Albiin et al. 2008), the SCS method was applied to diagnose cancer in a cohort with primary sclerosing cholangitis (PSC) ± cancer by analysis of bile ex vivo. This is necessary because strictures of the biliary system are difficult to classify. The only successful treatment for PSC is liver transplant. If cancer is present in the system, the transplant will fail. Applying the above methods to a cohort of 45 specimens, overall accuracy of 89% was achieved. This study was conducted in collaboration with the Karolinska Hospital and Stockholm University.

7 The Future

It is clear from these, and many other published studies, that in combination with robust, accurate analytical methods high resolution NMR is a powerful tool for the classification of specimens. In the case of readily available specimens such as blood or urine, this approach suffices. However, when tissue biopsies are used, or fluids such as bile are obtained by endoscopic techniques, a degree of invasion is involved. In the future for the latter cases we must seek to obtain useful spectra by studies in vivo. The MRI instruments currently available for such purposes involve magnetic fields of 1.5–4.0 Tesla, compared to the high resolution instruments reaching as high as 22 Tesla. The separation between the component resonances in MR spectra is linearly dependent on the strength of the magnetic field, as is approximately the signal-to-noise ratio of a given MR spectrum. One is tempted to measure in vivo MR spectra at the highest field possible, but there are several impediments: ultra high field instruments of adequate bore size for human studies are prohibitively expensive; there is risk involved at high fields due to neural stimulation and deposition of radiofrequency power. Hence, these high field instruments are not likely to appear in hospitals due to cost concerns, installation problems (the magnets are huge), and safety concerns. What we must do is work very hard to optimize the resolution of spectra by careful deconvolution and to design the various instrument components for optimum performance. Useful spectra have been obtained at reasonable fields (1.5–3.0 Tesla) and the quality of the spectra is improving rapidly. Thus, we believe that 3.0 or 4.0 Tesla instruments will be able to yield spectra which will be adequate for the screening. An example is ^{1}HMRS of bile in gall bladder in vivo in a commercial 3 Tesla instrument obtained in several minutes (Mohajeri et al. 2010, unpublished data).

The future looks bright!

References

Albiin, N. et al. (2008) Detection of cholangiocarcinoma with magnetic resonance spectroscopy of bile in patients with and without primary sclerosing cholangitis. *Acta Radiologica* **49**: 855–862.

Bezabeh, T. et al. (2009) Detecting colorectal cancer by ^{1}H magnetic resonance spectroscopy of fecal extracts. *NMR Biomed.* **22**(6): 593–600.

Erikkson, L. et al. (2001) *Multi- and megavariate data analysis – principles and applications*. Umetrics AB, Umeå.

Kuncheva, L.I. (2004) *Combining instance classifiers – methods and algorithms*. Wiley, Hoboken, NJ.

Nikulin, A.E. et al. (1998) Near-optimal region selection for feature space reduction: novel preprocessing methods for classifying MR spectra. *NMR Biomed.* **11**: 209–217.

Somorjai, R.L. et al. (1995) Computerized consensus diagnosis: a classification strategy for the robust analysis of NMR spectra. I. Application to thyroid neoplasms. *Magn. Res. Med.* **33**: 257–263.

Somorjai, R.L. et al. (2004a) A data-driven, flexible machine learning strategy for the classification of biomedical data. *Artificial intelligence methods and tools for systems biology* (Chapter 5), W. Dubitzky and F. Azuaje, eds. Computational Biology Series, Vol. 5, Springer, Dordrecht, pp. 67–85.

Somorjai, R.L. et al. (2004b) Mapping high-dimensional data onto a relative distance plane – a novel, exact method for visualizing and characterizing high-dimensional instances. *J. Biomed. Inform.* **37**: 366–379.

Somorjai, R.L. (2009) Creating robust, reliable, clinically relevant classifiers from spectroscopic data. *Biophys. Rev.* **1**: 201–211.

The Emergence and Ozone Treatment Studies of Living Cells

Davor Pavuna, Božidar Pavelić, Ognjen Pavičević, Domagoj Prebeg, and Mario Zovak

Abstract Most processes in biomedicine can hardly be addressed within some physicists' reductionist views. More appropriate is the emergence approach, the process of complex pattern formations from simpler rules, that was introduced by biologists, and that has a natural place within bio-complexity. There is nothing that commands the system to form an emergent pattern, but instead the interactions of each part, to its immediate surroundings, causes a complex process which leads to some form of order. One such emergent process in re-balancing of tissue cells is the high-frequency bio oxidative treatment. Ozone therapy is a well established alternative and complementary treatment in most mainland EU countries and The European Cooperation of Medical Ozone Societies publishes guidelines on medical indications and contraindications of ozone and hosts expert training seminars. Here we describe the accumulated know-how on the use of feeble active flux of ozone in the cell healing and briefly discuss several tested ozone treatments in dentistry and medicine (herpes, muscle recovery, injury recovery etc.).

Keywords Bio-physics · Emergence · Ozone · Dentistry · Herpes · Biomedicine

1 Introduction

It is appropriate that this chapter is dedicated to the memory of Greta Pifat-Mrzljak (Pavuna 2007) as she already organized more than half a dozen of summer schools before a complementary conference series "From Solid State to BioPhysics" has begun in Dubrovnik (Forró and Pavuna 2000). Over the years in numerous lectures, one of us (DPV) had the opportunity to hear the insights of many distinguished speakers in biophysics, biomedicine and related topics. And they all showed that we are still in the early days of science and especially in our understanding of

D. Pavuna (✉)
Physics Section, EPFL, CH-1015 Lausanne, Switzerland
e-mail: davor.pavuna@epfl.ch

J. Brnjas-Kraljević, G. Pifat-Mrzljak (eds.), *Supramolecular Structure and Function 10*, DOI 10.1007/978-94-007-0893-8_8,

biomedical complexity (Vinter et al. 2010). Nevertheless, today we still have more questions than answers and one very useful approach is to follow up on our present understanding of the biophysics (of the phase diagram) of a living cell. (Sackmann, www.darpa.mil/dso/personnel/23_math_chall_b_mann.pdf).

In such a model, the cell is considered as a self-organized unit that has clearly defined functioning thermodynamic range of temperature, oxidation and hydration. Any departure from the optimal phase diagram range causes a misbalance and eventually an illness. Within such a simple framework did we initiate our systematic studies of high frequency (HF) ozone treatments in dentistry and biomedicine, assuming that in many cases the cells may locally require some additional oxygen to restore the phase diagram balance. Indeed, as we illustrate, our preliminary studies indicate that this is a reasonable working model, although more detailed (clinical) studies are needed to establish the full scientific insight into various treatments.

2 On High-Frequency Bio Oxidative Treatments (HF-BOT)

The ozone therapy is a well established alternative and complementary therapy in most mainland European countries. The European Cooperation of Medical Ozone Societies, founded in 1972, publishes guidelines on medical indications and contraindications of ozone and hosts training seminars. In the early 1980s a German survey and investigation into ozone therapy by the University clinic in Giessen and the Institute for Medical Statistics, published in the Empirical Medical Acts revealed over 5 million ozone treatments had been delivered to some 350,000 patients, by more than 1,000 therapists, of this number about half were medical doctors (Jacobs 1982). Although ozone is used in a complementary capacity by a significant number of practitioners, it has still not gained popular support with mainstream biomedicine, partly as various health insurance schemes invoked hostile objections from pharmaceutical industry. In order to bring some light into the field we have begun systematic studies on the ozone therapies and here we describe the use of contemporary Tesla ozone generators and some of the ongoing studies of its use in biomedical cell treatments. We describe the accumulated know-how on the benefit of the feeble active flux of ozone in the cell healing and we critically discuss several tested ozone treatments in dentistry and medicine (acne, muscle recovery, cellulite, injury recovery etc.) (http://www.biozon.hr).

Given aforementioned delicate balance of temperature, water and oxygen within the living cell phase diagram, it is understandable that for the normal healthy person, the best way to increase cell-oxygen levels is to eat plenty of fresh fruit and vegetables and to avoid polluted air and cigarette smoke. Drinking spring water can also help, since the processes used to purify tap water also reduce its oxygen content. Assuming no genetic anomalies, a balanced lifestyle including some moderate aerobic exercise, fresh air, a healthy diet and sufficient rest seems to be the key to maintaining health and prolonging it. However, as we all know, most people at some point develop some health anomaly and in many cases high-frequency bio-oxidative therapy (HF-BOT) may help. The key hypothesis behind the reactive oxygen species (ROS) therapy, introduced by Otto Warburg, double Nobel Prize laureate is the statement that the human health is primarily dependant on oxygen level in the cells. (http://nobelprize.org/nobel_prizes/medicine/laureates/1931/warburg-bio.html).

The High-Frequency Bio Oxidative therapy (HF-BOT) is a therapy which uses the HF-BO generator to produce reactive oxygen species (ROS) in the treated area. Biozon generators, used in this study, are the first such generators in the world certified as medical devices, and were developed and tested for the past 15 years. The fundamentals of this kind of generator have been developed in 1896 by Nikola Tesla in USA and the know-how was gradually spread around the world. Due to the extremely low current of the generator, this kind of therapy is painless, safe and easy to use. The generator system consists of a central unit which produces high frequency electrical impulses with very low current, the applicator which induces the high voltage and the special plasma probe which is a conductor of the HF ozone to a desired area (see Figs. 1 and 2).

Fig. 1 The Reactive Oxygen Species (ROS) is produced from atmospheric oxygen within the tiny gap between the plasma probe and the tissue by the high frequency electrical field generated from the Tesla generator. www.biozon.hr

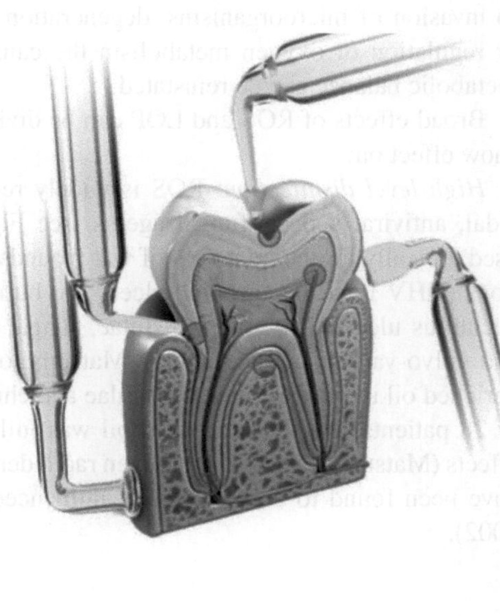

Fig. 2 In dentistry the application of local HF ozone application offers safe and accurate treatment. www.biozon.hr

There are four main effects by which the effect of HF-BOT can be explained: (a) Production of ROS in the treated region; (b) Electro-Stimulation; (c) Photo-Stimulation, and (d) Diathermy.

The Reactive Oxygen Species (ROS) is produced by high frequency electrical field which is formed on the tip of the Biozon plasma probe at the moment when probe closes the circuit with the human body. This HF field resonates in the frequency of Oxygen molecular bond and causes its breakage. This reaction produces several reactive oxygen forms known as ROS – nascent oxygen, O_3 and H_2O_2 – depending on the media – air or liquid. In human body ROS is transformed in LOP (Lipid per-Oxidation Products) when it gets in contact with PUFAs (Poly Unsaturated Fatty Acids). LOPs are very heterogeneous in nature so their effects are very diverse and affect many biochemical pathways and cells. Here we will not enter into discussions of all the possibilities.

Our preliminary insights can be crudely summarized as follows: The HF ozone has better "penetration" to reach the areas/organs that are not easily accessible. Eventually ozone seems to act locally in chemical terms just as the ordinary oxygen would. The oxygenation revives the cell, as water and temperature are more easily brought into the body-cell balance. The biophysics description evolves from the aforementioned cell-phase-diagram.

With the high level disinfection, which is known and used for more than century, today the ROS is recognized to be a powerful medical drug. Multiple medical effects of ROS have been researched in the past 100 years and there are now numerous studies which have shown that it certainly is one of the most powerful treatments due to its multiple beneficial effects on human body and cells. The key hypothesis behind the ROS therapy is aforementioned Warburg's insight is the statement that the human health is primarily dependant on oxygen level in the cells (http://nobelprize.org/nobel_prizes/medicine/laureates/1931/warburg-bio.html). Hypoxia (lowered concentration of oxygen in the tissue) is defined as the prime cause or complication in almost every disease. Hypoxic tissue is subject to invasion of microorganisms, degeneration or mutation. By delivery of oxygen or regulation of oxygen metabolism the cause of disease can be eliminated and metabolic balance can be reinstated.

Broad effects of ROS and LOP can be divided into several main groups as they show effect on:

High level disinfection: ROS is widely recognized as one of the best bactericidal, antiviral and antifungal agents (see Fig. 3). The ROS enriched oil is now used topically for the treatment of war wounds, anaerobic infections, herpetic infections (HHV I and II), atrophic ulcers and burns, cellulitis, abscesses, anal fissures, decubitus ulcers (bed sores), fistulae, fungal diseases, furunculosis and gingivitis and vulvo-vaginitis (Bocci 2005). Matsumoto et al. tested the efficacy of the ROS enriched oil in the treatment of fistulae and chronic surgical wounds and, in a series of 28 patients; the ROS enriched oil was fully effective in 27 cases without side-effects (Matsumoto et al. 2001). Even radio dermatitis lesions in patients with cancer have been found to be beneficially influenced by exposure to ROS (Jordan et al. 2002).

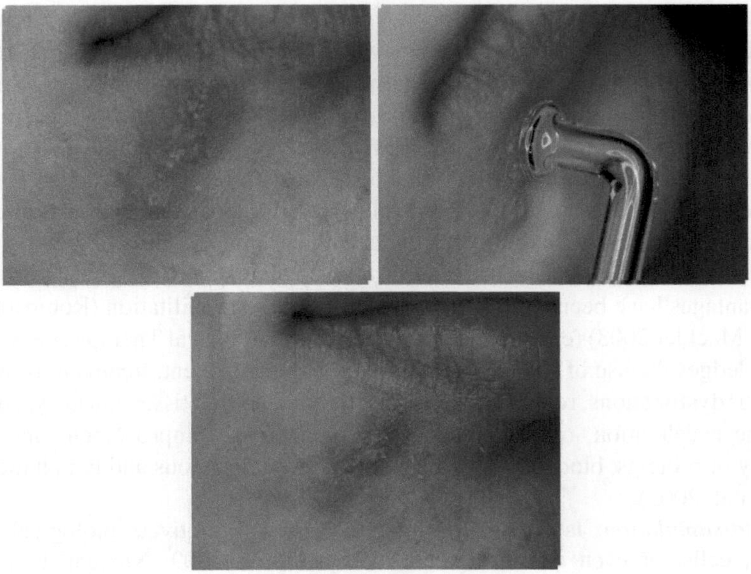

Fig. 3 HF ozone treatment is often very useful in pathology. Here we illustrate a successful treatment of herpes lesion with plasma ozone probe. Before (*left*) treatment (*right*) and 16 h after treatment (*below*)

Plasma: Oxidation of ascorbic acid, uric acid, and $(SH)^-$ groups of proteins and glycol-proteins and other molecules that play role in antioxidant response in body (Halliwell 1996). In the hydrophilic plasma environment LOP and H_2O_2 are generated in reaction between PUFAs and ROS (Pryor et al. 1995).

Erythrocytes – improved oxygenation in the body: Glycolysis and pentose pathway are accelerated with a consequent increase in ATP levels (Rokitansky 1982). Also, shift to the right of the oxygen-hemoglobin dissociation curve takes place in the re-infused erythrocytes due to slight decrease of intracellular pH or/and increase of 2,3-diphosphoglycerate (2,3 DPG) levels (Rokitansky 1982). That could be one of the key effects that are responsible for ROS having such a role in treating ischemia.

Leukocytes – improved immunological response: ROS seems to stimulate phagocytic activity of neutrophils and levels of immunoglobulins (Paulesu et al. 1991). ROS can also induce cytokine release from peripheral blood mononuclear cells (PBMC) (Bocci et al. 1993). For some interleukins (IL-8) it is well known that they are regulated by some reactive oxygen species like H_2O_2 (DeForge et al. 1992).

Platelets – growth factors, wound healing: ROS can induce significant release of platelet-derived growth factor (PDGF), transforming growth factor b1 (TGF-b1) and interleukin-8 (IL-8). These findings may explain the enhanced healing of ulcers in patients with chronic limb ischemia treated with ROS (Bocci et al. 1999).

Endothelium – vasodilatation: One member of ROS family, namely NO· has been recognized to be one of the crucial factors which regulate relaxation of blood vessels (Moncada 1992). During the de-oxygenating process NO· is liberated from this site and reacts with small thiols, resulting in S-nitroso-thiols with GSH or cysteine that have half-life of several hours that can slowly distribute NO· to endothelial receptors (Jia et al. 1996). These findings show that because of slow releasing of NO· not only its toxicity is impaired but also a steady relaxation of blood vessels is ensured. This may clarify good results in treating patients with ischemic tissues.

Electrical stimulation: The use of electrotherapy has been widely researched and the advantages have been well accepted in the field of rehabilitation (Robinson and Snyder-Mackler 2008) (electrical muscle stimulation) Physical Therapy Association acknowledges the use of Electrotherapy for: pain management, treatment of neuromuscular dysfunctions, restoration of connective and dermal tissue integrity, absorption rate acceleration, of the blood vessel permeability improvement, increases mobility of proteins, blood cells, inducement of arterial, venous and lymphatic flow (Alon et al. 2005).

Photo-stimulation: Is the use of light to artificially activate biological compounds, cells, or even whole organism (Rojas et al. 2009, Xuejuan and Xing 2009).

Diathermy: This term means "electrically induced heat" and is commonly used for muscle relaxation. Diathermy is used in treatment of chronic arthritis, bursitis, fractures, gynaecological diseases, sinusitis, and other conditions (Walker et al. 2007, Robinson 1929, Benedet 1990).

3 Some Examples of Dental and Medical Effects of HFBot

Here we give some simple illustrations of the biomedical treatments.

In Fig. 4 we show an example of the ozone treatment baseocellular carcinoma – carcinoma of the skin. The 67 year old patient had skin carcinoma over 3×4 cm anterior side of left lower lag. Standard therapy with different creams and bandages was conducted for 6 months with no success. Patient was treated with ozone therapy three times for only 3 min. And the result is visible on the pictures attached. Patient did not show any remission 3 years after therapy.

Another striking example is a case of a 50 year old male patient with brain tumor. The patient was paralyzed and without consciousness after the surgical and radiation therapy. Following 2 months of regular ozone therapy all body functions were restored. At a control in hospital 3 months after the first ozone therapy when the patient walked into the practice, his own doctors were stating: "we do not know how but whatever you are doing is helping and you are recovering as the tumor is decreasing in size and all functions are restored". After 2 years patient is living normal life.

There are many such examples that clearly show the potential of the ozone treatment, yet further systematic clinical studies are needed to fully elucidate the processes that are taking place and to optimize the medical treatments.

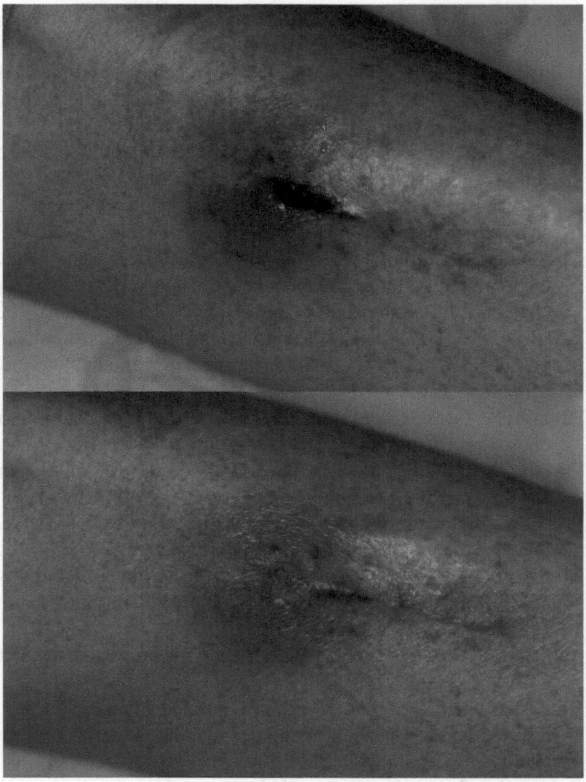

Fig. 4 Ozone treated baseocellular carcinoma – carcinoma of the skin. 3×4 cm anterior side of left lower lag. Standard therapy with different creams and bandages was conducted for 6 months with no success

Acknowledgements Davor Pavuna wants to acknowledge the support of the EPFL and Swiss NSF in Bern. Domagoj Prebeg acknowledges the support by Biozon R&D department.

References

Alon, G. et al. (2005) *Electrotherapeutic terminology in physical therapy: section on clinical electrophysiology*. American Physical Therapy Association, Alexandria, VA.

Benedet, J.L. (1990) Early cervical neoplasia: treatment methods. *Can. Fam. Physician* **36**: 945–947.

Bocci, V., Luzzi, E., Corradeschi, F., Paulesu, L., Rossi, R., Cardaioli, E. and Di Simplicio, P. (1993) Studies on the biological effects of ozone: 4. Cytokine production and glutathione levels in human erythrocytes. *J. Biol. Regulat. Homeost. Agent.* **7**: 133–138.

Bocci, V., Valacchi, G., Rossi, R., Giustarini, D., Paccagnini, E., Pucci, A.M. and Di Simplicio, P. (1999) Studies on the biological effects of ozone: 9. Effects of ozone on human platelets. *Platelets* **10**: 110–116.

Bocci, V. (2005) *Ozone: a new medical drug*. Springer, Dordrecht.

DeForge, L.E., Fantone, J.C., Kenney, J.S. and Remick, D.G. (1992) Oxygen radical scavengers selectively inhibit interleukin 8 production in human whole blood. *J. Clin. Invest.* **90**(5): 2123–2129.

Forró, L. and Pavuna, D. (2010) From Solid State to BioPhysics I–V. Regular conference series: http://dubrovnik2010.epfl.ch/

Gao, X. and Xing, D. (2009) Molecular mechanisms of cell proliferation induced by low power laser irradiation. *J. Biomed. Sci.* **16**(1): 4.

Halliwell, B. (1996) Antioxidants in human health and disease. *Annu. Rev. Nutr.* **16**: 33–50.

Jacobs, M.T. (1982) Untersuchung Über Zwischenfalle Und Typische Komplikationen In Der Ozon-Sauerstoff-Therapie. *Ozon Nachrichten* **1**: 5.

Jia, L., Bonaventura, C., Bonaventura, J. and Stamler, J.S. (1996) S-nitrosohaemoglobin: a dynamic activity of blood involved in vascular control. *Nature* **380**: 221–226.

Jordan, L., Beaver, K. and Foy, S. (2002) Ozone treatment for radiotherapy skin reactions: is there an evidence base for practice? *Eur. J. Oncol. Nurs.* **6**: 220–227.

Matsumoto, A., Sakurai, S. and Shinriki, N. (2001) Therapeutic effects of ozonized olive oil in the treatment of intractable fistula and wound after surgical operation. Proceedings of the 15th Ozone World Congress, London, UK, 11–15 September 2001, Medical Therapy Conference (IOA 2001, Ed). Speedprint MacMedia Ltd, Ealing, London, pp. 77–84.

Moncada, S. (1992) Nitric oxide gas: mediator, modulator, and pathophysiologic entity. *J. Lab. Clin. Med.* **120**: 187–191.

Paulesu, L., Luzzi, E. and Bocci, V. (1991) Studies on the biological effects of ozone: 2. Induction of tumor necrosis factor (TNF-a) on human leucocytes. *Lymphokine Cytokine Res.* **10**: 409–412.

Pavuna, D. (2007) From solid state to bio-complexity: on the emerging science of emergence. In: *Supramolecular structure and function 9*, G. Pifat-Mrzljak, ed. Springer, New York, NY, p. 273.

Pryor, W.A., Squadrito, G.L. and Friedman, M. (1995) The cascade mechanism to explain ozone toxicity: the role of lipid ozonation products. *Free Rad. Biol. Med.* **19**: 935–941.

Robinson, A.J. and Snyder-Mackler, L. (2008) Clinical electrophysiology: electrotherapy and electrophysiologic testing, 3rd ed. Lippincott Williams and Wilkins, Baltimore, MD, pp. 151–196, 198–237, 239–274.

Robinson, C.A. (1929) Treatment of pelvic inflammations by diathermy. *Proc. R. Soc. Med.* **22**(3): 339–348.

Rojas, J.C., Lee, J., John, J.M. and Gonzalez-Lima, F. (2009) Neuroprotective effects of near-infrared light in an in vivo model of mitochondrial optic neuropathy. *J. Neurosci.* (Author manuscript; available in PMC 10).

Rokitansky, O. (1982) Klinik und biochemie der ozon therapy. *Hospitals* **52**: 643–647.

Sackmann, E. http://www.cuso.ch/3e-cycle/physique/

Vinter, C. et al. (2010) *Nature*. New York Times.

Walker, N.A., Denegar, C.R. and Preische, J. (2007) Low-intensity pulsed ultrasound and pulsed electromagnetic field in the treatment of tibial fractures: a systematic review. *J. Athl. Train.* **42**(4): 530–535.

Toxicity Study of Nanofibers

Lenke Horváth, Arnaud Magrez, Beat Schwaller, and László Forró

Abstract The major contribution of nanotechnology to our life is the controlled synthesis of a large variety of nanofilaments (nanowires and nanotubes) which could be the basis of future devices. Although the expectations are large concerning the improvement of our everyday life due to nanostructures (sensors, vectors for therapies, photovoltaic devices, fast integrated circuits etc.), there is a growing fear related to their possible health hazards, strongly reminiscent to those of asbestos. We have studied 3 model nanofilaments: TiO_2 nanowires, carbon and boron nitride (BN) nanotubes using MTT assays. We tried to unravel the role of local catalytic activity, the importance of structural defects, functional groups and the tortuosity of these nanofilaments in their alteration of cell proliferation.

Keywords Health hazard · Nanofilaments · Carbon nanotubes · TiO_2 · Boron nitride nanotubes · Cell death · Local catalysis · Tortuosity

1 Introduction

The discovery of numerous man-made materials on the nanometer scale in the past decades has added a new dimension to the enormous progress in nanotechnology, which holds great potential in improving our quality of life. Engineered nanoparticles (NPs) with unprecedented physical and chemical properties such as carbon nanotubes (CNT), fullerenes or metal oxide-based particles are manufactured worldwide, and make ground-breaking impact on diverse science, engineering and commercial sectors. Along with the discovery of novel nanomaterials, the professional and public exposure to these materials will significantly increase in the forthcoming years. This exposure may be desired and voluntary as in the case of nanomedicine (e.g. drugs formulated as nanomaterials or implants produced from nanomaterials),

L. Horváth (✉)
Department of Medicine, Unit of Anatomy, University of Fribourg, 1700 Fribourg, Switzerland;
Laboratory of Physics of Complex Matter, Ecole Polytechnique Fédérale de Lausanne,
1015 Lausanne, Switzerland
e-mail: lenke.horvath@epfl.ch

J. Brnjas-Kraljević, G. Pifat-Mrzljak (eds.), *Supramolecular Structure and Function 10*, DOI 10.1007/978-94-007-0893-8_9,

but in other cases not. Based on the structural similarity of some of the newly developed nanomaterials with previously investigated ones known to pose a serious health problem, e.g. asbestos fibers, the potential risk of man-made nanomaterials need not to be underestimated. It is known that inhalation of asbestos fibers induces progressive fibrotic disease of the lung (asbestosis) and lung cancer. Furthermore, when translocated from lung into the pleural space, asbestos fibers can end up in the mesothelial cell lining of the pleura. After a prolonged incubation period (10–30 years) these can induce mesotheliomas, one of the most aggressive tumors. The precise mechanism of asbestos toxicity and tumorigenesis is not known, but the presumed important parameters are being intensively studied (shape, surface chemistry, defect points on surface of the mineral fibers) (Barrett et al. 1989, Jaurand et al. 2009). It is hoped that the studies of different types of nanofilaments, even if they are not severely toxic, can bring important insight into the process of asbestos toxicity.

The manufacturing of CNTs, inorganic nanofilaments (INFs: nanowires, nanotubes) and other nanostructures is very challenging. Many of these are foreseen to become building blocks of new technologies and are highly desirable in many industrial products such as flat panel screens, composite materials or catalysis supports, etc. To date relatively little is known about the potential adverse effects of these novel nanomaterials on human health and the environment. Their introduction into industry requires the evaluation of their safety and the understanding of the impact of nanomaterials on the environment, biological species and human health (Shvedova et al. 2009).

1.1 Carbon Nanotubes

After their discovery in 1991 (Iijima 1991), carbon nanotubes (CNTs) have aroused great interest among researchers working on nanomaterials due to their exceptional properties including small size, large surface area, superior strength, high electrical conductivity and effectiveness in thermal conduction.

CNT can be imaginatively produced by rolling up a single layer of graphene sheet (single-walled CNT) or by rolling up many layers to form concentric cylinders (multi-walled CNT) (Bianco et al. 2005). Significant progress has been made in using CNTs in various fields, e.g. electronics (Carey 2003, Robertson 2006) and imaging (Hafner et al. 2001). In particular, many applications that employ CNTs have been proposed including biosensors, drug and vaccine delivery vehicles and novel biocomposite materials (Kam et al. 2005, Wu et al. 2005, Lin et al. 2004).

1.2 Inorganic Nanofilaments

INFs can be produced from many materials ranging from elemental semiconductors, metal oxides, chalcogenides to pnictides (Rao and Nath 2003, Patzke et al. 2002, Rao et al. 2007). This allows the synthesis of nanofilaments with well-defined and customer-desired properties (size, geometry, aspect ratio, electronic properties, etc.).

Significant progress has been made in incorporating TiO_2 nanoparticles in numerous applications for several decades. For example, nanostructured TiO_2 is the most popular catalyst for air and water purification systems, based on the photodegradation of organic pollutants (Mills et al. 1993); glasses or textiles are rendered self-cleaning when coated with TiO_2 nanoparticles (Gueneau-Rancurel 2007). It is widely used as food- and pharmaceutical additives (Lomer et al. 2002) as well as in cosmetic products, e.g. sunscreens contain nanostructured TiO_2, which efficiently filters ultraviolet radiation (Dransfield 2000). Last but not least, large-scale arrays of TiO_2-based nanofilaments (including nanotubes and nanowires) are used in photovoltaic cells and in photoelectrolyzer for the production of hydrogen by water splitting (Shankar et al. 2008). Due of this high popularity of TiO_2 nanostructures, it is important to study their effect on living matter.

1.3 Boron Nitride Nanotubes

Boron nitride nanotubes (BNNT) represent an important class of non-carbon inorganic nanotubes. Structurally, they are very close analogues of CNTs, where C atoms are fully substituted by B and N atoms, in a graphitic like sheet with almost no change in atomic spacing (Chopra et al. 1995), but actually possess superior properties, displaying far better thermal and chemical stabilities than CNTs. BNNTs withstand thermal treatment in air up to 900°C without deterioration (Golberg et al. 2001), whereas CNTs are burned at 400–500°C. Thus, the BNNT utilization is much more feasible as far as high temperature and/or reactive environments are concerned. They may be used as protective capsules for any type of nanomaterials or devices, which otherwise would not be stable in air, and/or easily contaminated at ambient conditions (Golberg et al. 2010). They provide an alternative choice to CNTs for mechanical reinforcement and improving thermal conductivity in polymeric composites (Wang et al. 2010). To date BNNTs have attracted much less attention than their carbon counterparts. This is attributed primarily to difficulties in BNNTs synthesis using well-established techniques for standard CNT fabrication, e.g. arc-discharge, laser ablation or chemical vapour deposition (CVD). It is expected that with the elaboration of a high yield synthesis of BNNT, there will be a high demand for applications. It will be very useful to know ahead of time, the health hazards related to their manipulation.

2 Current State of the Reserach

It has been shown that nanomaterials can enter the human body by several pathways, and most importantly this includes the airways, the digestive tract and the skin, which are in constant contact with the environment. Due to their small size, nanomaterials can translocate from these entry portals into the blood circulation and lymphatic systems, and ultimately to body tissues and organs. Eventually, nanomaterials may be distributed within the whole body (Hoet et al. 2004).

Consequently, the risk of nanomaterials for different organs needs to be assessed. Beforehand, toxicity studies should be carried out at a lower level of complexity, on the level of cells. The mechanisms in which various nanomaterials exert their adverse effects may be different for different cell types and may depend on the properties of the numerous types of nanomaterials. They can produce irreversible damage to cells by different mechanisms including oxidative stress and/or organelle injury as well as an impairment of the cell cycle. Finally, the exposure to nanomaterials may have short-term effects (acute toxicity), but may also lead to chronic toxicity or long-term negative effects such as carcinogenesis.

2.1 Potential Toxicity of CNTs

The toxicity of CNTs has been an important issue in the scientific community because CNTs have fiber-like characteristics in terms of their elongated shape, dimensions and aspect ratio. Thus, in many ways they resemble asbestos fibers. Considerable efforts have been made to investigate the toxicity and biocompatibility of CNTs both in vitro and in vivo (Smart et al. 2006, Jain et al. 2007, Poland et al. 2008). CNTs have been tested in vitro in various cell types e.g. lung epithelial cells, mesothelial cells, keratinocytes, fibroblasts, and some authors have demonstrated that CNTs induce cytotoxicity and/or inflammatory responses (Ye et al. 2009, Tabet et al. 2009, Simon-Deckers et al. 2008, Pacurari et al. 2008, Magrez et al. 2006, Shvedova et al. 2003), while others have found CNTs to be biocompatible when cultured with cells in vitro (Shi et al. 2008, Chlopek et al. 2006). The data concerning the potential hazards related to CNT exposure to date is not conclusive. This apparent inconsistency in the literature might be due to the fact that different CNTs were used in these studies. It is known that (I) carbon nanotubes are synthesized and purified by different methods, (II) they can be single- or multi-walled, (III) they have various length and diameter distribution and (IV) can be subjected to numerous surface modifications. Furthermore, often different cell lines and different assays were employed for the toxicological evaluation of these materials. These variations, not only for carbon nanotubes, but also for nanoparticles in general, present significant challenges in the assessment of their potential toxicity (Card et al. 2008).

2.2 Potential Toxicity of INFs

The experimental results concerning the potential adverse effects of nanostructured TiO_2 are also under debate (Nohynek et al. 2007). First, the dermal penetration of TiO_2 nanoparticles is discussed. Once entered in the vascular system, nanostructured TiO_2 could be distributed throughout the body to organs, various tissues and exposed cells (Fabian et al. 2008, Vileno et al. 2007). Second, while the toxicity of TiO_2 in its isotropic form (particles with diameter ranging from few nanometres to micrometers) has been widely studied (Lewinski et al. 2008, Warheit et al. 2007), the toxic action of TiO_2-based nanofilaments was essentially unknown until recently (Magrez et al. 2009).

2.3 *Potential Toxicity of BNNTs*

The least information on the toxicity of the above-discussed materials is available on BNNTs, possibly due to the marginal progress in their research compared to CNTs. The first biocompatibility study on BNNTs was carried out in 2008 by Ciofani et al., who investigated polyethyleneimine-coated BNNTs interaction with human neuroblastoma cells. They showed good cytocompatibility and cellular uptake by an energy-dependent endocytic process, indicating the suitability of these materials for biomedical applications. Another study showed that pure BNNTs, i.e. without any dispersing agent, were not cytotoxic to human embryonic kidney cells (HEK 293) when compared to MWCNTs, suggesting their use in therapeutic or diagnostic applications (Chen et al. 2009). Although BNNTs seem to be non-toxic, the controversy on the toxicological issues seen in case of CNTs necessitates reconfirmation under varying experimental conditions, by extending in vitro investigations to relevant cell types, which would present primary biological targets of direct exposure to these nanomaterials.

3 Cellular Toxicity

It should be emphasised that not all NPs produce adverse health effects. Accumulating evidence in the literature shows that parameters influencing the toxicity of nanoparticles include: length, shape, surface structure, surface chemistry, aggregation state (i.e. presence of a dispersing agent) and contamination of metal catalyst and amorphous carbon (Simon-Deckers et al. 2008). In this work we will review our findings that: (I) surface chemistry of the nanofilaments does matter; (II) that surface defects promote the local catalysis thus inducing toxicity; (III) and that tortuosity, i.e. the amount of twists and turns may influence the internalisation of nanofilaments and consequently, these parameters are likely to have a role in the toxicity.

3.1 *Effect of Surface Chemistry*

Our research group has previously focused on the influence of surface chemistry of multiwalled carbon nanotubes (MWCNTs) and carbon nanofibers (CNFs) on acute cell toxicity (Magrez et al. 2006). The results are summarized below.

3.1.1 Materials and Methods

Synthesis of nanofibers and preparation of exposure solutions: Multiwalled carbon nanotubes (MWCNTs) were produced by chemical vapor deposition of acetylene (C_2H_2) over Fe_2Co supported by $CaCO_3$ at 660°C. As a purification process, both MWCNTs and CNFs materials were dispersed in 1 M HCl overnight after incubation for 15 min in a sonication bath. After filtration, the materials were washed with distilled water and dried at 150°C overnight. Both materials were further annealed at high temperature to remove any chemical groups bonded to the surface (1,000°C for 2 h under dynamic vacuum). To achieve surface filament modifications, purified MWCNTs and CNFs were dispersed in 1 M nitric acid overnight.

Gelatine (porcine skin Type A, Sigma) was used to stabilize the carbon-based nanomaterials (CBN) suspension of particles and prevent aggregation. Excess gelatine was removed by washing CBNs in distilled water. We could not find any indication of CBNs functionalisation. Gelatine was further added to cell media to yield identical gelatine concentrations in all samples. All CBN solutions were prepared as stock solutions containing 20 μg/ml CBN and 20 μg/ml gelatine and stock CBN solutions were prepared by sonication. After sterilization of CBN solutions in an autoclave at 121°C for 15–20 min, the solutions were diluted to concentrations of 0.2, 0.02 and 0.002 μg/ml for cellular toxicity experiments.

Characterization of test particles: MWCNTs (Fig. 1a) with an average diameter of 20 nm and aspect ratios ranging from 80 to 90 and CNFs (Fig. 1b, obtained from Pyrograf Products, Inc.) with a mean diameter of 150 nm and aspect ratios of 30–40 were employed in the study.

Cell culture and in vitro cytotoxicity measurements: The cytotoxicity assay was performed on human lung tumour cells (American Type Culture Collection) NCI H596 (Adenosquamous carcinoma). The cells were incubated with gelatine-containing medium (control) and CBNs in gelatine-containing medium. The cytotoxicity of the CBNs was evaluated by the MTT (3-(4,5 dimethylthiazol-2-yl)-2,5-diphenyltetrazolium bromide) assay (Thiazolyl blue tetrazolium bromide (M5655); Sigma). The assay is based on the accumulation of dark blue formazan crystals inside living cells after their exposure to MTT. The destruction of the cell membrane by the addition of dimethylsulfoxide (DMSO) results in the liberation and solubilisation of the crystals. The number of viable cells is thus directly proportional to the level of the initial formazan product created. The formazan concentration is finally quantified using a spectrophotometer by measuring the

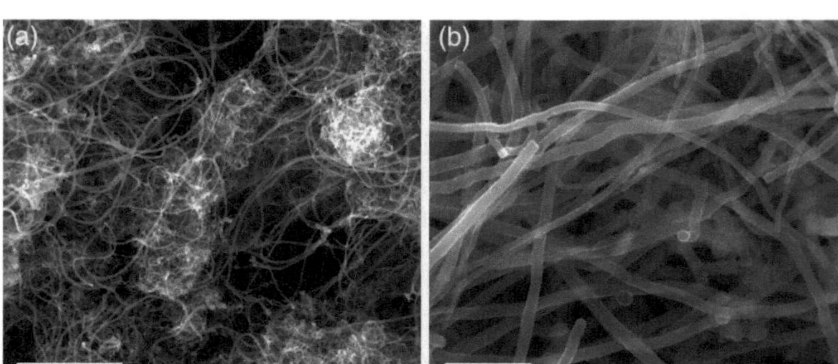

Fig. 1 Scanning electron microscopy images of (**a**) MWCNTs, (**b**) CNFs. The aspect ratio of these nanoparticles is about 80–90 and 30–40, respectively. The scale bars correspond to 2 μm (taken from Magrez et al. 2006)

absorbance at 540 nm (ELISA reader). For each cell type a linear relationship between cell number and optical density was established, thus allowing an accurate quantification of changes in the rate of cell proliferation.

Results: To explore the effect of chemically active surface sites on the toxicity, a set of experiments was performed in which the surface chemistry of the filaments was modified. MWCNTs and CNFs were decorated with a chemical modification of the outer layer of the CNT resulting in carbonyl (C=O), carboxyl (COOH), and/or hydroxyl (OH) groups attached onto the nanotube and nanofiber surfaces. H596 cells were grown in a gelatine-containing medium containing 0.02 μg/ml of dispersed CBNs. Cells grown in a plain gelatine-containing medium served as reference. The MTT assay was carried out between days 1 and 4 after exposure to CBNs. The chemical decoration effect on cytotoxicity is displayed in Fig. 2. The toxicity increased with the chemical surface treatment, which was most significant in the case of MWCNTs and moderate for CNFs. These results clearly demonstrated that grafting additional putatively "toxic" chemical groups on the surface of MWCNTs reduced the number of viable cells significantly.

3.2 Effect of Morphology and Surface Chemistry

In our recently published study (Magrez et al. 2009) we investigated the acute cytotoxicity of TiO_2-based nanofilaments on H596 lung tumour cells in vitro in relation to their morphology and surface chemistry. The experiments were performed on epithelial cells, based on the fact that lung epithelial cells are one of the first cell types to encounter the nanosized materials (like e.g. TiO_2 nanofilaments) released into the environment.

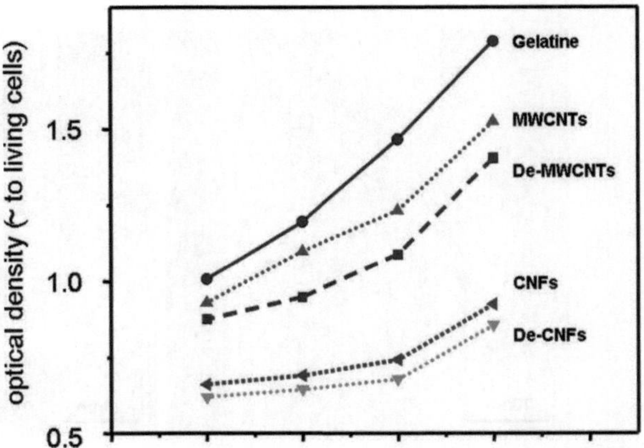

Fig. 2 Effect of filament decoration on cell toxicity in H596 cells. The growth curves obtained from chemically decorated MWCNTs and CNFs are denoted De-MWCNTs and De-CNFs, respectively. The filament concentration to which all samples were exposed to was 0.02 μg/ml. In both cases, the number of viable cells is lower in the decorated samples, indicative of increased toxicity (taken from Magrez et al. 2006)

3.2.1 Materials and Methods

Synthesis of nanofilaments and preparation of exposure solutions: The synthesis of the nanofilaments was performed by hydrothermal treatment. Nanotubes were prepared by adding 6 g of anatase (Alfa Aesar, 99.9%) to 28 ml of a 15 M NaOH solution (Aldrich, 99.99%). The mixture was subsequently heated to 150°C for 72 h. For the synthesis of nanofilaments, a 10 M NaOH solution was used and the temperature was raised to 170°C, while the amount of anatase and the duration of hydrothermal treatment remained identical. The white precipitate produced was washed with distilled water and dried at 120°C overnight. The ionic exchange was performed by dispersing, for 2 h, the produced nanofilaments in diluted HCl with H/Na ratio equal to 100.

The TiO_2-based nanofilaments were dispersed in a highly diluted gelatine solution to minimize aggregation as described previously (Magrez et al. 2006). Gelatine containing suspensions of TiO_2-based nanofilaments were stable for more than a month. SEM analysis of nanowires deposited by the boil deposition process onto Si wafers revealed that the nanofilaments were dispersed individually (Lee et al. 2007). In addition, some of the experiments were repeated using Tween 80 (Sigma-Aldrich, Buchs, Switzerland) as the dispersing substance, a nondenaturing detergent previously used in several toxicity studies of nanomaterials (Wick et al. 2007).

Characterization of test particles: Both TiO_2 nanotubes (Fig. 3a) or TiO_2 nanowires (Fig. 3b) can be obtained by tuning the growth conditions (namely, temperature and composition of the starting mixture Ti/Na ratio). Unlike carbon nanotubes, the TiO_2-based nanotubes are not perfect cylinders. They are better described as a scroll of a TiO_2 layer. These TiO_2-based multiwalled nanotubes exhibit an average diameter of around 12 nm with approximately five walls.

Cell culture and in vitro cytotoxicity measurements: The cytotoxic effect of TiO_2-based nanofilaments on NCI H596 cells was evaluated by MTT assay as described previously (Magrez et al. 2006). In addition direct counting of H596 cells was

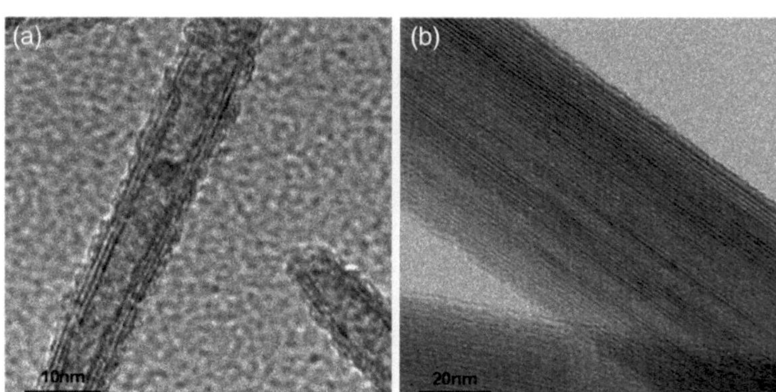

Fig. 3 Transmission electron microscopy images of TiO_2-based nanotubes (**a**) and nanowires (**b**) (modified from Magrez et al. 2009)

performed. Cells were seeded on round laminin-coated glass covers, exposed to nanomaterials (2 μg/ml) for 4 days, fixed, and stained with HE (hematoxylin and eosin). Microphotographs (20× magnification) of the stained cells were acquired by a systematic random approach similar to the one described before (Maetzler et al. 2004). Briefly, a region of the round glass cover slips close to the periphery was randomly selected, and on the acquired image, the numbers of cells were counted. On the basis of this initial starting point, the other three regions of the glass cover slips to be analyzed were determined by a predefined L-shaped scheme. This prevented a bias toward selecting particular regions on the glass cover slips. The acquired images were randomly numbered and analyzed by the experimenter not knowing the key until the final calculations. From each glass cover, the number of cells (determined by counting the dark stained nuclei) on each of the four images were added, yielding one value per glass cover slip.

Results: An average growth curve of untreated H596 cells (control) for 4 days is shown in Fig. 4a. Cells treated with TiO_2-based nanofilaments (2 μg/ml) were impaired in their proliferation/cell metabolic activity which could be detected by a decrease in the MTT signal, and was the most pronounced after 4 days (Fig. 4a). To compare the cytotoxicity results on TiO_2-based nanofilaments with the previous results obtained with carbon nanotubes (Magrez et al. 2006), multiwalled carbon nanotubes (MWCNTs) were selected, which were displaying an intermediate toxicity. In each nanofilament-treated sample the effect was more pronounced as the nanofilament concentration increased from 0.02 to 2 μg/ml (Fig. 4b). $Na_xTiO_{2+\delta}$ nanowires had the least effect on cell proliferation/activity, therefore they were considered to be the least toxic at all concentrations and at time points later than 1 day of exposure. The treatment by hydrochloric acid performed on the $Na_xTiO_{2+\delta}$ nanofilaments strongly enhanced the toxic action of the TiO_2-based nanofilaments. Particularly, at high nanofilament concentrations (2 μg/ml), the cell number/viability decreased in the following sequence: $Na_xTiO_{2+\delta}$ NWs > MWCNTs > $H_yTiO_{2+\delta}$ NWs ≈ $H_yTiO_{2+\delta}$ NTs. In the protonated form, there was no clear difference between TiO_2-based nanotubes as compared to the nanowires with respect to the MTT signal. This could indicate that the surface chemistry of the nanofilaments had a greater effect upon the cell proliferation/activity than their morphological characteristics. As the MTT assay measures the combined effects of cell proliferation and metabolic activity of cells and was reported to be prone to artefacts under certain experimental conditions (Worle-Knirsch et al. 2006), results presented in Figs. 4a and 5 were validated by another method, i.e. directly counting the number of surviving cells from microphotographs. For this, the time point (day 4) and nanomaterial concentration (2 μg/ml) was selected, where differences between the various nanomaterials were most evident. In comparison to untreated cells, both the MTT signals (ANOVA: $p < 0.0001$) and number of cells (ANOVA: $p < 0.0001$) were decreased in all nanomaterial-treated samples (Fig. 4c). According to post-hoc analysis by the Tukey-test, the samples can be grouped into three classes (white, gray and black bars in Fig. 4c). Samples with the same shading in Fig. 4c1 were significantly different from the other groups (all at $p < 0.05$). The only exception was the Na^+ form of the nanotubes (NaT

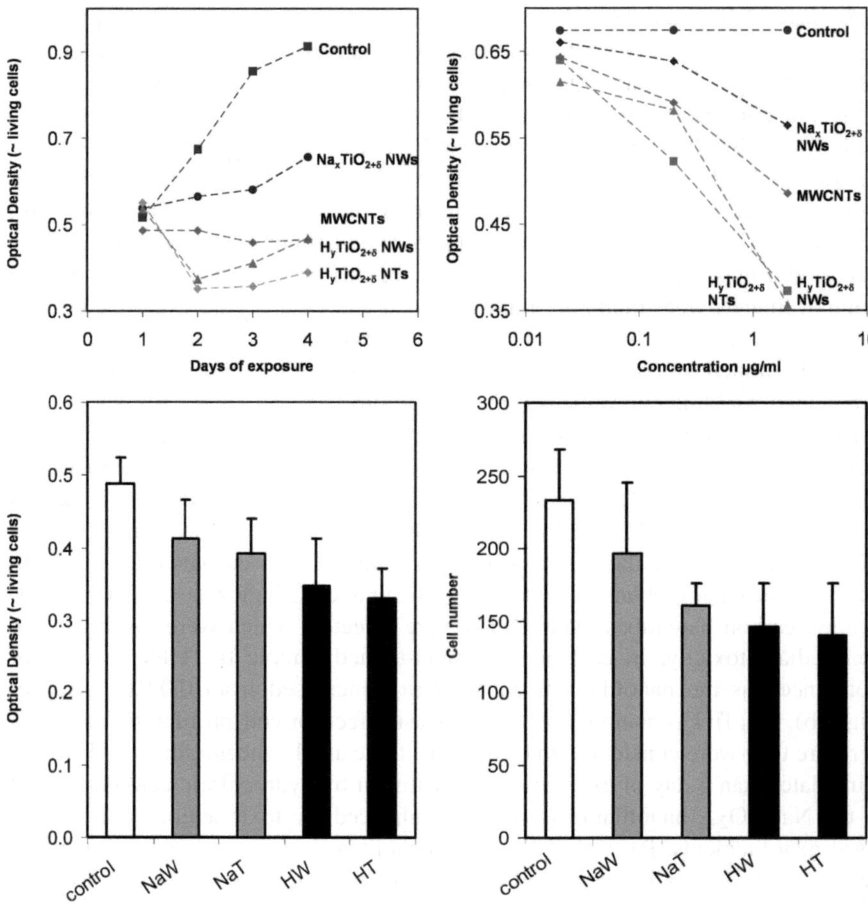

Fig. 4 (**a**) Representative growth curve for H596 cells grown in normal medium (control), or 2 µg/ml of MWCNTs, $Na_xTiO_{2+\delta}$, and $H_yTiO_{2+\delta}$ nanofilaments (NWs, nanowires; NTs, nanotubes). (**b**) Dose-dependent toxicity of H596 cells exposed to the tested materials for 2 days. (**c**) Cytotoxicity of TiO_2-based nanomaterials determined by MTT assays (**c1**) and cell counting (**c2**). H596 cells were seeded in 96-well plates for MTT assays or on laminin-coated coverslips for cell counting and were exposed to nanomaterials for 4 days. The numbers in **c1** and **c2** represent averages ± standard deviations from four independent experiments. In the MTT assay (**c1**), each sample was measured in triplicate, and in **c2**, the cell number was determined from four microphotographs per sample. Results were analyzed by ANOVA, followed by posthoc analyses by a Tukey test (NaW and NaT, nanowires and nanotubes in the Na^+ form (*gray bars*); HW and HT, nanowires and nanotubes in the H^+ form (*black bars*)) (taken from Magrez et al. 2009)

in Fig. 4c2) which was not significantly different from the group of the nanofilaments in the H^+ form. However, the toxicity order, i.e. that the Na^+ forms were less toxic than the H^+ forms was also seen in the direct cell counting. In summary, the two methods, MTT assay and cell counting yielded essentially identical results.

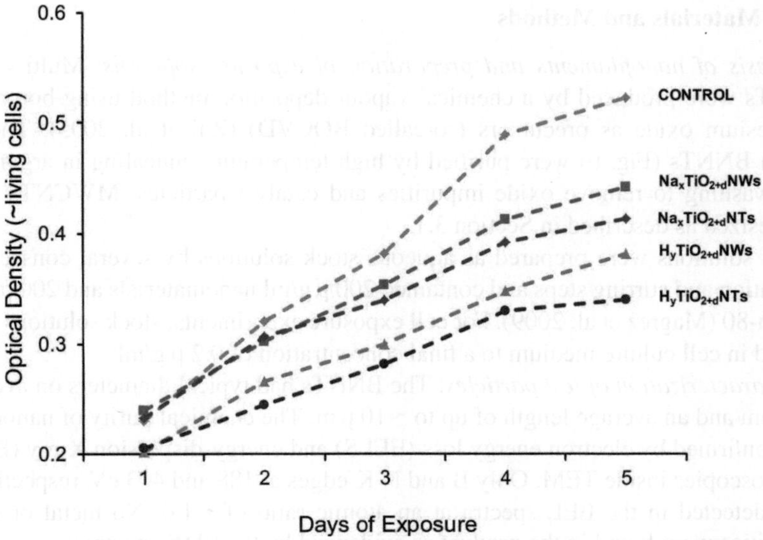

Fig. 5 Growth curve of H596 cells in the presence of TiO_2-based nanotubes and nanowires in the Na^+ and H^+ forms (2 μg/ml). For these experiments, Tween-80 was used as the dispersing agent. A comparison with the experiments obtained in gelatine-containing medium (Fig. 4a) revealed the results to be essentially identical, i.e. the toxicity order was as follows: $Na_xTiO_{2+\delta}$ NWs < $H_yTiO_{2+\delta}$ NWs ≈ $H_yTiO_{2+\delta}$ NTs. In addition, also $Na_xTiO_{2+\delta}$ nanotubes (NT) were tested. As in Fig. 4a the HCl treated nanofilaments were more toxic than the Na^+ forms and the nanotubes were slightly more toxic than the nanowires. However, the differences between tubes and wires were smaller than differences between HCl-treated (H^+) and Na^+ forms. Tween-80 was used as the dispersing agent, since Tween-80 containing solutions can be stored for prolonged periods (>1 year). In gelatine-containing solutions the proteinaceous parts start to decompose resulting in nonspecific growth-inhibiting effects seen in the MTT assays when comparing old (> 6 months) with freshly prepared (< 1 month) gelatine-containing solution, even in the absence of nanofilaments (data not shown) (taken from supplemental information of Magrez et al. 2009)

Our investigation shows that geometry appears to play a role, i.e. the thinner nanotubes both in the H^+ form (Fig. 4a, c) and in the Na^+ form (Fig. 5) were slightly more toxic than the corresponding nanowires. However, surface chemistry affected the survival of exposed cells more importantly. In the case of $Na_xTiO_{2+\delta}$ when Na^+ is exchanged for a smaller H^+ ion, the structural imperfections left-behind act as chemically active sites, and render the nanostructures more toxic.

3.3 Possible Effect of Tortuosity

In our most recent study, the toxicity of BNNTs was evaluated in A549 epithelial cells and RAW 264.7 macrophages. Besides epithelial cells, as mentioned previously, alveolar macrophages are the first line of defence against nanomaterials upon inhalation. In addition to BNNTs, cells were exposed to pristine and functionalized MWCNTs as well and the viability was followed with MTT-assay.

3.3.1 Materials and Methods

Synthesis of nanofilaments and preparation of exposure solutions: Multi-walled BNNTs were produced by a chemical vapour deposition method using boron and magnesium oxide as precursors (so-called BOCVD) (Zhi et al. 2005). The as-grown BNNTs (Fig. 6) were purified by high temperature annealing in argon and acid washing to remove oxide impurities and catalyst particles. MWCNTs were synthesized as described in Section 3.1.

All solutions were prepared as aqueous stock solutions by several consecutive sonication and stirring steps and contained 200 μg/ml nanomaterials and 200 μg/ml Tween-80 (Magrez et al. 2009). For cell exposure experiments, stock solutions were diluted in cell culture medium to a final concentration of 0.2 μg/ml.

Characterization of test particles: The BNNTs had typical diameters on average ~50 nm and an average length of up to ~10 μm. The chemical purity of nanotubes was confirmed by electron energy loss (EELS) and energy-dispersion X-ray (EDX) spectroscopies inside TEM. Only B and N K-edges at 188 and 403 eV, respectively, were detected in the EEL spectra at an atomic ratio of ~1.0. No metal or other impurities were found in the product as evidenced by the EDX spectra.

Cell culture, in vitro cytotoxicity measurements, light microscopical imaging: for the experiments the following cell lines were used: A549 human type II lung epithelium cells, often used for lung toxicity assays (Tabet et al. 2009, Pulskamp et al. 2007) and RAW 264.7 mouse macrophage cells, and the cell proliferation was evaluated by the MTT (3-(4,5-dimethylthiazol-2-yl)-2,5-diphenyltetrazolium bromide) assay as described before (Magrez et al. 2006, 2009). For the cytopathological examination RAW 264.7 (1.4×10^4 cells/well) were seeded in 12-well culture plates on laminin-pretreated glass cover slips, and were further cultured either in

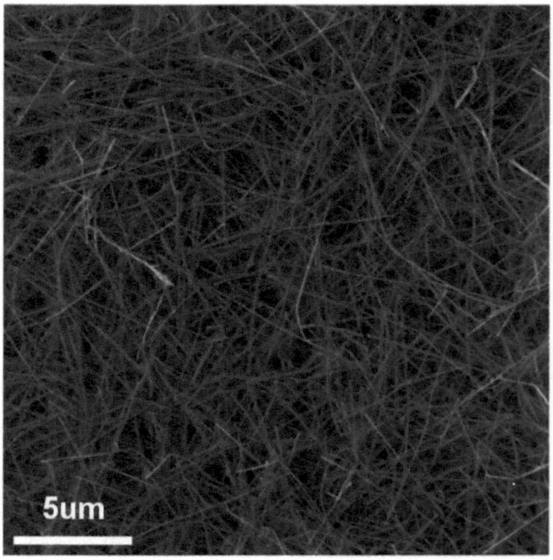

Fig. 6 Scanning electron microscopy image of multi-walled BNNTs

medium containing 2 μg/ml Tween 80 (control) or in medium containing 2 μg/ml nanomaterials. Four days post exposure the remaining cells were fixed with ice-cold (–20°C) acetone-methanol (1:1) solution for 10 min and stained with standard Hematoxyline-Eosine (HE) staining solutions.

Results: Treatment of A549 epithelial cells and RAW 264.7 macrophages with nanomaterials (NMs) for 5 days revealed a time dependent toxicity (Fig. 7a, b). Average growth curves of A549, and RAW 264.7 cells exposed to nanomaterials (0.2 μg/ml) showed a decreased MTT signal compared to untreated cells (control). The toxic action of all materials on the two cell types was already apparent on the second day post exposure and the effect was even more pronounced after culturing the cells for 5 days. The MTT signals were profoundly smaller in all NM-treated samples. In general, both cell types were more susceptible to BNNTs than to MWCNTs. Furthermore, at each time point later than 2 days, the pristine MWCNTs (p-MWCNTs) were less toxic than functionalized MWCNTs (f-MWCNTs), which is in line with our previous report on the cytotoxicity of MWCNTs determined in other lung epithelial cells (Magrez et al. 2006). The enhanced toxicity of BNNTs can be explained with the fact that these NMs have lower tortuosity than their CNT counterparts. Thus, this straighter and needle-shaped structure may facilitate their penetration in the cells and interfere with cell organelles and intracellular biochemical processes. The cytotoxicity is even more pronounced in the case of cells, which are specialized to remove foreign material from an organism, i.e. the macrophages (Horváth et al. 2010). These cells are activated upon contact with nanomaterials, usually engulfing these by phagocytosis. Sometimes they fail to efficiently engulf their target (the NMs), and are likely to undergo "frustrated phagocytosis", stimulating further inflammation and cell damage. Figure 8 depicts typical morphological alterations of RAW 264.7 macrophages frequently observed in BNNT treated samples, e.g. formation of multinucleated giant cells and shrinkage of cells with nuclear condensation after 4 day of 2 μg/ml BNNT exposure.

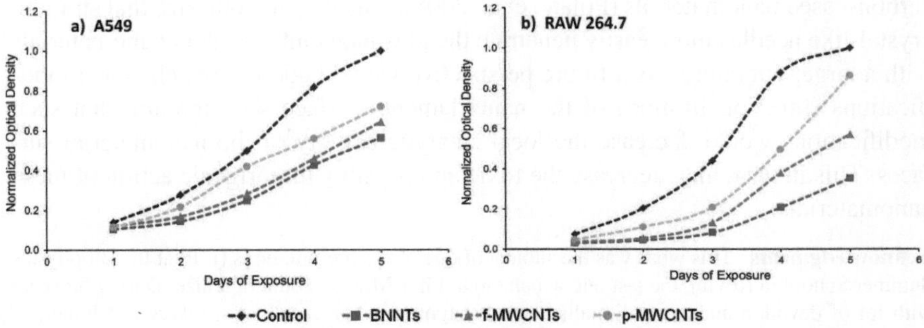

Fig. 7 (**a, b**) Normalized time-dependent toxicity of A549 and RAW 264.7, cells treated with 0.2 μg/ml of BNNT, p-MWCNT and f-MWCNT for 5 days determined by MTT assay

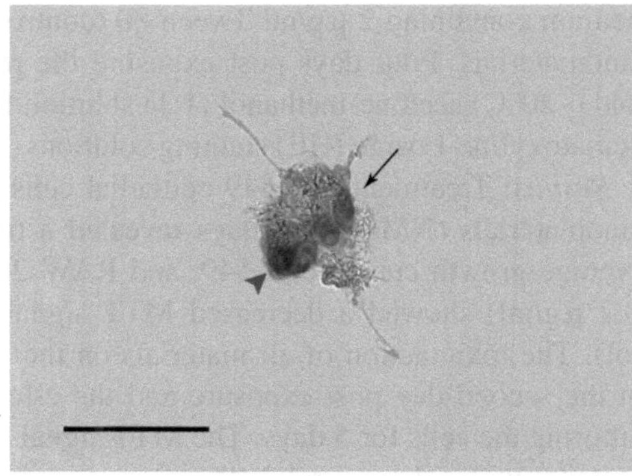

Fig. 8 Typical morphological alterations after treatment with 2 μg/ml BNNTs in RAW 264.7 cells. A multinucleated giant cell engulfed large amounts of nanofibers (*arrow*) and is in close apposition to a strongly stained eosinophilic cell with a shrunken (*condensed*) nucleus (*arrowhead*). Scale bar: 20 μm

4 Conclusions and Future Perspectives

One important aspect of nanomaterial toxicity is the shape of the nanomaterials, even when it is composed of the same atoms (e.g. carbon) (Poland et al. 2008, Magrez et al. 2006). Furthermore, toxicity depends on the cell lines; three different tumour cell lines, all derived from lung epithelial cells show quantitative differences in the degree of toxicity (Magrez et al. 2006). Another aspect is the surface modification of nanomaterials. The carbon-based nanofibers were more toxic in the decorated form (Magrez et al. 2006). Also the ionic exchange from Na^+ to H^+ in the TiO_2-based nanomaterials (nanowires and nanotubes) results in structural point defects as well as extended defects within the nanofilament crystal structures. These structural changes on the surfaces also enhanced the nanomaterial's toxicity. We had hypothesized that these structural imperfections may act as sites for catalytic processes such as production of radicals or interaction sites for proteins that might lead to conformational changes of bound protein affecting the normal cell physiology. The results from the studies on BNNTs indicate that also the tortuosity of the nanofilaments may affect nanomaterials toxicity, as previously shown also for carbon-based nanomaterials (Poland et al. 2008). One may hypothesize that straight, crystal-like needles more easily penetrate the plasma membrane than nanomaterials with a larger curvature. As a future perspective we will address the chemical modifications (functionalisation) of the nanofilaments surface with the aim that such modifications would decrease the local catalytic activity of the nanomaterial surfaces. This in turn, may decrease the toxic and possibly tumorigenic action of these nanomaterials.

Acknowledgments This work was the subject of one of my presentations (L.F) at the Biophysics Summer School in Rovinj, the last one which Greta Pifat-Mrzljak could organize. During 30 years with lot of devotion and professionalism she brought together excellent speakers and hundreds of young students. Under the Mediterranean sky she cultivated a very creative atmosphere. Her memory occupies a permanent place in our hearts.

This work is supported by the Swiss National Science Foundation. The supply of BN nanotubes by Dmitri Goldberg is gratefully acknowledged.

References

Barrett, J.C., Lamb, P.W. and Wiseman, R.W. (1989) Multiple mechanisms for the carcinogenic effects of asbestos and other mineral fibers. *Environ. Health Perspect.* **81**: 81–89.

Bianco, A., Kostarelos, K. and Prato, M. (2005) Applications of carbon nanotubes in drug delivery. *Curr. Opin. Chem. Biol.* **9**: 674–679.

Card, J.W., Zeldin, D.C., Bonner, J.C. and Nestmann, E.R. (2008) Pulmonary applications and toxicity of engineered nanoparticles. *Am. J. Physiol. Lung Cell Mol. Physiol.* **295**: L400–L411.

Carey, J.D. (2003) Engineering the next generation of large-area displays: prospects and pitfalls. *Philos. Trans. A Math. Phys. Eng. Sci.* **361**: 2891–2907.

Chen, X., Wu, P., Rousseas, M., Okawa, D., Gartner, Z. et al. (2009) Boron nitride nanotubes are noncytotoxic and can be functionalized for interaction with proteins and cells. *J. Am. Chem. Soc.* **131**(3): 890–891

Chlopek, J., Czajkowska, B., Szaraniec, B., Frackowiak, E., Szostak, K. and Beguin, F. (2006) In vitro studies of carbon nanotubes biocompatibility. *Carbon* **44**: 1106–1111.

Chopra, N.G., Luyken, R.J., Cherrey, K., Crespi, V.H., Cohen, M.L., Louie, S.G. and Zettl, A. (1995) Boron-nitride nanotubes. *Science* **269**: 966–967.

Ciofani, G., Raffa, V., Menciassi, A. and Cuschieri, A. (2008) Cytocompatibility, interactions, and uptake of polyethyleneimine-coated boron nitride nanotubes by living cells: confirmation of their potential for biomedical applications. *Biotechnol. Bioeng.* **101**: 850–858.

Dransfield, G.P. (2000) Inorganic sunscreens. *Radiat. Prot. Dosimetry* **91**: 271–273.

Fabian, E., Landsiedel, R., Ma-Hock, L., Wiench, K., Wohlleben, W. and van Ravenzwaay, B. (2008) Tissue distribution and toxicity of intravenously administered titanium dioxide nanoparticles in rats. *Arch. Toxicol.* **82**: 151–157.

Golberg, D., Bando, Y., Kurashima, K. and Sato, T. (2001) Synthesis and characterization of ropes made of BN multiwalled nanotubes. *Scr. Mater.* **44**: 1561–1565.

Golberg, D., Bando, Y., Huang, Y., Terao, T., Mitome, M., Tang, C. and Zhi, C. (2010) Boron nitride nanotubes and nanosheets. *Acs Nano* **4**: 2979–2993.

Gueneau-Rancurel, L. (2007) Chemistry for architecture: the self-cleaning glass. *Actual Chim.* **311**: 6–10.

Hafner, J.H., Cheung, C.L., Woolley, A.T. and Lieber, C.M. (2001) Structural and functional imaging with carbon nanotube AFM probes. *Prog. Biophys. Mol. Biol.* **77**: 73–110.

Hoet, P.H., Bruske-Hohlfeld, I. and Salata, O.V. (2004) Nanoparticles – known and unknown health risks. *J. Nanobiotechnol.* **2**: 12.

Horváth, L., Magrez, A., Forró, L. and Schwaller, B. (2010) Cell type dependence of carbon based nanomaterial toxicity. *Physica Status Solidi (B)*. **247**: 3059–3062.

Iijima, S. (1991) Helical microtubules of graphitic carbon. *Nature* **354**: 56–58.

Jain, A.K., Mehra, N.K., Lodhi, N., Dubey, V., Mishra, D.K., Jain, P.K. and Jain, N.K. (2007) Carbon nanotubes and their toxicity. *Nanotoxicology* **1**: 167–197 and references therein.

Jaurand, M.C.F., Renier, A. and Daubriac, J. (2009) Mesothelioma: do asbestos and carbon nanotubes pose the same health risk? *Part Fibre Toxicol.* **6**: 16.

Kam, N.W., O'Connell, M., Wisdom, J.A. and Dai, H. (2005) Carbon nanotubes as multifunctional biological transporters and near-infrared agents for selective cancer cell destruction. *Proc. Natl. Acad. Sci. USA* **102**: 11600–11605.

Lee, K., Duchamp, M., Kulik, G., Magrez, A., Seo, J.W., Jeney, S., Kulik, A.J. and Forró, L. (2007) Uniformly dispersed deposition of colloidal nanoparticles and nanowires by boiling. *Appl. Phys. Lett.* **91**: 173112.

Lewinski, N., Colvin, V., Drezek, R. (2008) Cytotoxicity of nanoparticles. *Small* **4**: 26–49.

Lin, Y., Taylor, S., Li, H.P., Fernando, K.A.S., Qu, L.W., Wang, W., Gu, L., Zhou, B. and Sun, Y.P. (2004) Advances toward bioapplications of carbon nanotubes. *J. Mater. Chem.* **14**: 527–541.

Lomer, M.C.E., Thompson, R.P.H. and Powell, J.J. (2002) Fine and ultrafine particles of the diet: influence on the mucosal immune response and association with Crohn's disease. *Proc. Nutr. Soc.* **61**: 123–130.

Maetzler, W., Nitsch, C., Bendfeldt, K., Racay, P., Vollenweider, F. and Schwaller, B. (2004) Ectopic parvalbumin expression in mouse forebrain neurons increases excitotoxic injury provoked by ibotenic acid injection into the striatum. *Exp. Neurol.* **186**: 78–88.

Magrez, A., Kasas, S., Salicio, V., Pasquier, N., Seo, J.W., Celio, M., Catsicas, S., Schwaller, B. and Forró, L. (2006) Cellular toxicity of carbon-based nanomaterials. *Nano Lett.* **6**: 1121–1125.

Magrez, A., Horvath, L., Smajda, R., Salicio, V., Pasquier, N., Forró, L. and Schwaller, B. (2009) Cellular toxicity of TiO_2-based nanofilaments. *Acs Nano* **3**: 2274–2280.

Mills, A., Davies, R.H. and Worsley, D. (1993) Water-purification by semiconductor photocatalysis. *Chem. Soc. Rev.* **22**: 417–425.

Nohynek, G.J., Lademann, J., Ribaud, C. and Roberts, M.S. (2007) Grey goo on the skin? Nanotechnology, cosmetic and sunscreen safety. *Crit. Rev. Toxicol.* **37**: 251–277.

Pacurari, M., Yin, X.J., Ding, M., Leonard, S.S., Schwegler-Berry, D., Ducatman, B.S., Chirila, M., Endo, M., Castranova, V. and Vallyathan, V. (2008) Oxidative and molecular interactions of multi-wall carbon nanotubes (MWCNT) in normal and malignant human mesothelial cells. *Nanotoxicology* **2**: 155–170.

Patzke, G.R., Krumeich, F. and Nesper, R. (2002) Oxidic nanotubes and nanorods – anisotropic modules for a future nanotechnology. *Angew. Chem. Int. Ed. Engl.* **41**: 2446–2461.

Poland, C.A., Duffin, R., Kinloch, I., Maynard, A., Wallace, W.A.H., Seaton, A., Stone, V., Brown, S., MacNee, W. and Donaldson, K. (2008) Carbon nanotubes introduced into the abdominal cavity of mice show asbestos-like pathogenicity in a pilot study. *Nat. Nanotechnol.* **3**: 423–428.

Pulskamp, K., Diabate, S. and Krug, H.F. (2007) Carbon nanotubes show no sign of acute toxicity but induce intracellular reactive oxygen species in dependence on contaminants. *Toxicol. Lett.* **168**: 58–74.

Rao, C.N.R. and Nath, M. (2003) Inorganic nanotubes. *Dalton Trans.* **1**: 1–24.

Rao, C.N.R., Vivekchand, S.R.C., Biswasa, K. and Govindaraja, A. (2007) Synthesis of inorganic nanomaterials. *Dalton Trans.* **34**: 3728–3749.

Robertson, J. (2006) Growth of nanotubes for electronics. *Mater. Today* **10**: 36–43.

Shankar, K., Bandara, J., Paulose, M., Wietasch, H., Varghese, O.K., Mor, G., LaTempa, T.J., Thelakkat, M. and Grimes, C.A. (2008) Highly efficient solar cells using TiO_2 nanotube arrays sensitized with a donor-antenna dye. *Nano Lett.* **8**: 1654–1659.

Shi, X.F., Sitharaman, B., Pham, Q.P., Spicer, P.P., Hudson, J.L., Wilson, L.J., Tour, J.M., Raphael, R.M. and Mikos, A.G. (2008) In vitro cytotoxicity of single-walled carbon nanotube/biodegradable polymer nanocomposites. *J. Biomed. Mater. Res. A* **86A**: 813–823.

Shvedova, A.A., Castranova, V., Kisin, E.R., Schwegler-Berry, D., Murray, A.R., Gandelsman, V.Z., Maynard, A. and Baron, P. (2003) Exposure to carbon nanotube material: assessment of nanotube cytotoxicity using human keratinocyte cells. *J. Toxicol. Environ. Health A* **66**: 1909–1926.

Shvedova, A.A., Kisin, E.R., Porter, D., Schulte, P., Kagan, V.E., Fadeel, B. and Castranova, V. (2009) Mechanisms of pulmonary toxicity and medical applications of carbon nanotubes: two faces of Janus? *Pharmacol. Ther.* **121**: 192–204.

Simon-Deckers, A., Gouget, B., Mayne-L'hermite, M., Herlin-Boime, N., Reynaud, C. and Carriere, M. (2008) In vitro investigation of oxide nanoparticle and carbon nanotube toxicity and intracellular accumulation in A549 human pneumocytes. *Toxicology* **253**: 137–146.

Smart, S.K., Cassady, A.I., Lu, G.Q. and Martin, D.J. (2006) The biocompatibility of carbon nanotubes. *Carbon* **44**: 1034–1047.

Tabet, L., Bussy, C., Amara, N., Setyan, A., Grodet, A., Rossi, M.J., Pairon, J.C., Boczkowski, J. and Lanone, S. (2009) Adverse effects of industrial multiwalled carbon nanotubes on human pulmonary cells. *J. Toxicol. Environ. Health A* **72**: 60–73.

Vileno, B., Lekka, M., Sienkiewicz, A., Jeney, S., Stoessel, G., Lekki, J., Forró, L. and Stachura, Z. (2007) Stiffness alterations of single cells induced by UV in the presence of nanoTiO_2. Environ. Sci. Technol.Stiffness alterations of single cells induced by UV in the presence of nanoTiO_2. *Environ. Sci. Technol* **41**: 5149–5153.

Wang, J., Lee, C.H. and Yap, Y.K. (2010) Recent advancements in boron nitride nanotubes. *Nanoscale* **2**: 2028–2034.

Warheit, D.B., Hoke, R.A., Finlay, C., Donner, E.M., Reed, K.L. and Sayes, C.M. (2007) Development of a base set of toxicity tests using ultrafine TiO_2 particles as a component of nanoparticle risk management. *Toxicol. Lett.* **171**: 99–110.

Wick, P., Manser, P., Limbach, L.K., Dettlaff-Weglikowska, U., Krumeich, F., Roth, S., Stark, W.J. and Bruinink, A. (2007) The degree and kind of agglomeration affect carbon nanotube cytotoxicity. *Toxicol. Lett.* **168**: 121–131.

Worle-Knirsch, J.M., Pulskamp, K. and Krug, H.F. (2006) Oops they did it again! Carbon nanotubes hoax scientists in viability assays. *Nano Lett.* **6**: 1261–1268.

Wu, W., Wieckowski, S., Pastorin, G., Benincasa, M., Klumpp, C., Briand, J.P., Gennaro, R., Prato, M. and Bianco, A. (2005) Targeted delivery of amphotericin B to cells by using functionalized carbon nanotubes. *Angew. Chem. Int. Ed. Engl.* **44**: 6358–6362.

Ye, S.F., Wu, Y.H., Hou, Z.Q. and Zhang, Q.Q. (2009) ROS and NF-kappa B are involved in upregulation of IL-8 in A549 cells exposed to multi-walled carbon nanotubes. *Biochem. Biophys. Res. Commun.* **379**: 643–648.

Zhi, C.Y., Bando, Y., Tan, C.C. and Golberg, D. (2005) Effective precursor for high yield synthesis of pure BN nanotubes. *Solid State Commun.* **135**: 67–70.

Subject Index

J. Brnjas-Kraljević, G. Pifat-Mrzljak (eds.), *Supramolecular Structure and Function 10*, DOI 10.1007/978-94-007-0893-8,

MIX
Papier aus verantwortungsvollen Quellen
Paper from responsible sources
FSC® C105338

If you have any concerns about our products,
you can contact us on
ProductSafety@springernature.com

In case Publisher is established outside the EU,
the EU authorized representative is:
Springer Nature Customer Service Center GmbH
Europaplatz 3, 69115 Heidelberg, Germany

Printed by Libri Plureos GmbH
in Hamburg, Germany